Imitation in Human and Animal Behavior

Imitation in Human and Animal Behavior

Wanda Wyrwicka

TRANSACTION PUBLISHERS
New Brunswick (U.S.A.) and London (U.K.)

Copyright © 1996 by Transaction Publishers, New Brunswick, New Jersey 08903.

All rights reserved under International and Pan-American Copyright Conventions. No part of this book may be reproduced or transmitted in any form or by any means, electronic or mechanical, including photocopy, recording, or any information storage and retrieval system, without prior permission in writing from the publisher. All inquiries should be addressed to Transaction Publishers, Rutgers—The State University, New Brunswick, New Jersey 08903.

This book is printed on acid-free paper that meets the American National Standard for Permanence of Paper for Printed Library Materials.

Library of Congress Catalog Number: 95-21634
ISBN: 1-56000-246-8
Printed in the United States of America

Library of Congress Cataloging-in-Publication Data

Wyrwicka, Wanda
   Imitation in human and animal behavior / Wanda Wyrwicka.
      p. cm.
   Includes bibliographical references and index.
   ISBN 1-56000-246-8 (alk. paper)
      1. Imitation in children.   2. Visual learning.   3. Learning in animals.
   4. Imitation—Social aspects.   5. Psychology, Comparative.   I. Title.
   BF723.I53W87   1995
   156'.31523—dc20   95-21634
                                                                          CIP

Contents

Acknowledgments

I wish to thank my colleagues, Drs. Michael H. Chase, Carmine D. Clemente, Ronald M. Harper, Rebecca K. Harper, and Barry M. Sterman, for reading parts of the manuscript and helpful comments.

I also wish to give my thanks to Dr. Yojiro Kawamura, Dr. Andrew N. Meltzoff and Dr. David P. Phillips, as well as the editors and publishers of *American Journal of Sociology, Behaviour,* and *Science* for permission to reproduce some of the illustrations used in this book. I also wish to thank Dr. David Shaffer and his research assistant, Mr. Roger C. Hicks for assistance in providing literature related to imitation of suicide.

Special thanks are due to my daughter, Joanna-Veronika Warwick for her valuable editorial assistance.

Introduction

The topic of this book is an often observed form of behavior which recently began to attract the attention of a number of researchers. This behavior, called the "imitative behavior" or simply "imitation" (the term first used by Tarde 1890), can roughly be described as the copying by an individual of a certain motor or vocal act performed by another individual (usually of the same species).

In the past, many cases of imitation in both humans and animals, including imitative acts in newborn human babies, imitation by following the companion, development of imitation skills later in life, and learning by imitation, were reported by various researchers. Imitative behavior was also demonstrated in the process of development of complex functions of the organism such as feeding and food preferences. The suppression of imitative behavior in some conditions was also observed.

However, the large amount of observational and experimental data related to the problem of imitation is dispersed in many separate studies, and only rarely discussed in review articles, especially those related to behavioral neuroscience. There is a need for a study in which the neural mechanisms of imitation can be explored and general rules of imitative behavior established. This volume attempts to fulfill, at least partially, this need. It should be stressed, however, that this book is not a comprehensive review of all studies on imitation; rather it concentrates on selected cases of imitative behavior.

The book is composed of eleven chapters. The first five chapters describes the results of studies on humans. The first chapter includes the observations on imitative abilities in human neonates demonstrated at about one hour after birth. The second chapter relates the imitative acts in growing children below three years of age. The third chapter describes selected cases of imitative behavior in children over three years of age.

The topic of the fourth chapter is the role of imitation in cognitive development of children and adolescents. This chapter also describes the use of imitation as a method in therapy of phobias. Finally, the fifth chapter concentrates on imitation related to the tragic social problem of suicide among adolescent and adults, citing statistical and clinical data.

Chapters 6 to 9 present the data obtained mostly in studies on feeding in animals. The sixth chapter describes the results of observations on the influence of a companion (parent, sibling, or specific model) on development of independent feeding in birds and mammals. An example of interspecies imitation in feeding is also given. An addendum to this chapter describes some cases of vocal imitation in birds.

The seventh chapter elaborates on the facilitatory role of the presence of the mother in initiation of eating new food by young kittens while the eighth chapter demonstrates the imitation of the mother's improper food selection by weanling kittens. The ninth chapter presents some observations on learning by imitation in rats, cats, and monkeys and also offers some data related to learning by following the leader.

The tenth chapter describes the cases of inhibition of imitation in both humans and animals. It also includes a clinical case of disinhibition of imitative behavior after ablations of prefrontal lobes of the brain.

The concluding eleventh chapter summarizes the observational and experimental data, and then discusses the possible neural mechanism of imitative behavior. The theoretical approaches of several authors are described, and the hypothetic brain mechanism responsible for imitative behavior is proposed.

1

Imitative Behavior in Human Neonates

Imitative acts in infants have been reported by a number of investigators. Even as early as in 1900, Preyer described imitation of adults' lip protrusion by his 4-month-old child, an observation confirmed by McDougall (1908) who on several occasions saw that his 4-month-old child repeatedly protruded his tongue when an adult (whose face the child had been watching) made this movement.

Systematic research of the problem of infant imitative behavior was undertaken seventy years later. Studies by Meltzoff and Moore (1977) showed that imitation in neonates can occur much earlier than previously reported. Specifically, 12 to 21-day-old infants were found to be already able to imitate facial expressions of an adult, such as mouth opening, tongue protrusion and lip protrusion (figure 1.1). The infants were also able to imitate manual gestures, such as hand closing and opening when these gestures were performed in the front of them by an adult. These experiments were carefully designed; the experimenters made sure that the infants had not been previously trained to imitate any of these gestures. To prevent the possibility of such training, the parents, who were present during the test, had not been told in advance about the experimental plan. Each facial or manual gesture was demonstrated to the infant by the adult four times in a 15-second presentation period. Each presentation was followed by an interval of 20 seconds, during which the behavior of the infant was closely observed and videotaped. The videotapes were then examined and evaluated by six independent volunteer judges who confirmed that imi-

3

FIGURE 1.1

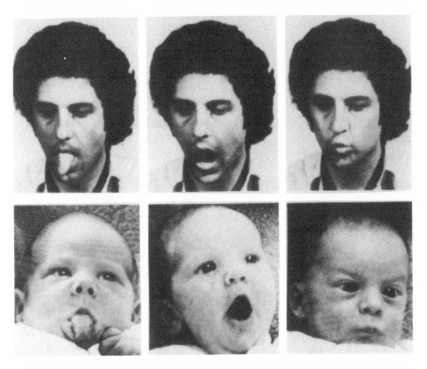

Sample photographs from video tape recordings of 2-to 3-week-old babies imitating (a) tongue protrusion; (b) mouth-opening; and (c) lip protrusion demonstrated by an adult experimenter.

Source: Reproduced from A.N. Meltzoff and M.K. Moore "Imitation of facial and manual gestures by human neonates," *Science*, 198: 75-78, 1977, by courtesy of the authors and permission from the American Association for Advancement of Science. Copyright 1977 by the AAAS.

tation took place, that is, the infants did perform the movements that had been demonstrated by the adult.

More recently Meltzoff and Moore (1983) tested their findings on neonates of less than 72 hour of age with low illumination observations using infrared-sensitive video equipment. The video recordings were then analyzed by an observer who was not informed about gestures shown to the infants. The results showed that facial gestures such as mouth opening and tongue protrusion were indeed imitated by neonates as young as 0.7 to 71 hours of age.

Several other investigators also reported cases of imitation of facial gestures in babies below four months of age (Dunkeld 1979) and in 2 to 10-week-old infants (Burd and Milewski 1981).

Field et al. (1982) showed that not only facial gestures but also specific facial expressions, such as happiness, sadness and surprise, presented by an adult model, can also be imitated by neonates. The study was conducted on 74 neonates at a mean 36 hours of age. During the test, the distance between the face of the adult model and the face of the infant was about 25 cm. The facial expressions of the neonate were recorded by an observer who was standing behind the model, and could see only the infant's face but not the face of the model. There were three series of trials, one for each facial expression. Each trial was performed first when the infant looked away from the model's face for about two seconds. An analysis of records showed that the neonates indeed imitated the model's facial expressions such as a happy smile, sad protrusion of lips, and mouth opening suggesting surprise.

Imitation of tongue protrusion in 2 to 6-week-old infants as a response to demonstration of this act by an adult, was also reported by Jacobson (1979). The research of this author included a test as to whether tongue protrusion could also be elicited by other stimuli. Observations were conducted on 24 infants at 6, 10, and 14 weeks of age. It was found that a black pen or a white ball moving in the front of the infant also produced tongue protrusion. However, matching behavior to the tongue model declined by 12 weeks of age.

Other tests conducted by Jacobson (1979) were related to the response of hand opening and closing. It appeared that these responses could be elicited in 14-week-old infants either by hand opening and closing or by a dangling ring. A comparative analysis of the data showed that the stimuli from the human model produced higher mean number of tongue protrusions or hand closing and opening than did stimuli such as pen or ball. Moreover, some investigators reported that they were unable to replicate the data obtained by Jacobson (Burd and Milewski 1981, quoted by Field et al. 1982). According to Field et al. (1982), certain differences in experimental procedure could be responsible for this failure. Nevertheless, the observations made by Jacobson (1979) that stimuli other than those deriving from living models can also produce imitative responses are important for understanding mechanisms involved in the imitation process. These mechanisms will be discussed in theoretical comments in chapter 11.

2

Imitation in Growing Children Under Three Years of Age

Imitative behavior in young children attracted the attention of the number of scientists early in this century. The classic studies on this topic are those of McDougall (1908), Guillaume (1926), and Piaget (1962). All these authors described a variety of cases of imitation and presented their theories of imitative behavior. Because most of the experimental data were provided by Piaget (1962), his work will be described in more detail below.

Piaget's observations on imitative behavior made on babies and children of various age led him to postulate that the ability to imitate develops together with the individual experience acquired by the child with age. On the basis of observations made mostly on his own children, he found that there exist six stages of development of imitative behavior. Let us describe each of these stages.

Stage I: Preparation Through the Reflex

This stage includes the period of the first few days after birth. At that time the experience of the baby is very limited, and the neonate is able to imitate only the most primitive motor acts (such as those described in chapter 1). For instance, Piaget observed that his child T., being drowsy but not actually asleep, began to cry when other babies in the same room began to wail. A few minutes later, when the baby's father tried to imitate the baby's interrupted whimpering, T. started to cry in earnest; other sounds, such a whistle, did not produce any reaction. The author concluded that "in neither case is there imitation but merely the starting off of a reflex by an external stimulus" (p. 7).

Stage II: Sporadic Imitation

This period includes a period from one to five months, during which the child's experience broadens by incorporating certain new visual or acoustic elements. This results in an addition of new motor or vocal responses. For instance, in the case of suction, the baby may use new gestures, such as putting the thumb into the mouth. Several cases of vocal imitation by babies in the second month of age were also observed. For example, baby T., being awake, motionless, and silent, started to cry three times in succession, when an older child was crying; each time the baby stopped crying as soon as the older child stopped. It was also observed that the young babies more readily repeated the specific sounds produced by an adult when the adult first repeated the sound produced by the baby. Thus, when the baby made sounds such as "la," "le," and so on, and the adult reproduced them, the baby repeated these sounds seven times out of nine. Similar observations were also made on another child.

The same was found concerning motor responses. They were reproduced by the child more easily when the adult first repeated the baby's motor act. However, it was found that the movement to be imitated had to be clearly visible to the baby. In the case when the movement was not distinctly visible, the imitation did not take place.

Stage III: The Beginning of Systematic Imitation

This stage includes the period between six (or earlier in some cases) and eight months of age. The observations made by Piaget at that time comprise instances of imitation of sounds belonging to the phonation of the child, and the imitation of movements the child has already made and seen. A number of cases are described in which the babies imitated the adult's sounds such as "pfs," "bva," "hha," and others. Other observations include the imitation of movements such as closing and opening the hand, when the adult repeated them several times. However, some movements of the mouth connected with eating, or movements of individual fingers, were not imitated. According to the author, these observations confirm the earlier hypothesis of Guillaume (1926) that "at this stage there is no spontaneous imitation of movements which the child cannot see himself make" (Piaget 1962, p. 29). For instance, the child can see the movements of the mouth when they are made by adults during eating, but is only aware of them in himself through kinesthetic and gustative sensations. Piaget concludes that "imitation of known sounds and visible movements proved to be lasting after a few mutual imitations, whereas imitation of nonvisible movements would have required, for its consolidation, a succession of sanctions alien to immediate assimilation" (p. 29).

Stage IV, Part 1: Imitation of Movements not Visible on the Body of the Subject

The first part of stage IV comprises the imitative behavior of the child between eighth and ninth month of age. At that time, the child is able "to assimilate the movements of others to those of his own body, even when his own movements are not visible to him" (p. 30). During this period, similarly as during the second and the third stages, the observations included mutual imitation (in which the adult first imitated a spontaneous facial movement of the child, thus producing the repetition of the same movement by the child). In a case when the baby put out her tongue and said "ba...ba," the adult quickly imitated her, and she began laughing; after several repetitions of this demonstration, without any sounds by the adult, the child "moved her lips and bit them for a moments and then put out her tongue several times in succession without making any sound" (p. 31). In another case, the adult rubbed his eyes in front of child J. just after she had rubbed her eye; after several repetitions of rubbing the eyes by the adult, the child imitated this movement each time. Similar observations were made with putting a finger into the ear.

Stage IV, Part 2: Beginning of Imitation of New Auditory and Visual Models

This part describes the imitative behavior of babies between 9 and 11 months of age. However, according to the author, it is "only the fifth stage that a general method of imitation of what is new can be developed" (p. 52).

Stage V: Systematic Imitation of New Models Including the Movements Invisible to the Child

Observations at this stage were made on children from one year and one year and 3 to 4 months of age (although a few cases refer to somewhat younger or older subjects). Let us describe several examples of imitation occurring at this stage.

Imitation of visible movements includes a case when the model (father) put a sheet of paper in front of child J. (1 year and 21 days of age) and made a few pencil strokes on it and then put down the pencil. The child seized the pencil and first made several unsuccessful movements, such as trying to draw with the left hand or using the wrong end of the pencil. Finally, after the model made marks on the paper with the finger, the child at once imitated him with her finger.

Further examples of imitation at this stage include the movements connected with various parts of the body which the child could see but with which he or she is not very familiar. (1) When the model rubbed his thigh with his right hand, child J watched him, laughed, and then rubbed first her cheek and then her chest. (2) One month later, when the model struck his abdomen, the child hit the table and then her own knees. Over three months later, when the model rubbed his stomach, the child hit first her knees, and then her thigh. It was only after a further repetition of this movement by the model, two weeks later (when the child was already one year and 4½ months old), that she correctly imitated the rubbing of the stomach by the model.

Other examples show the cases of child's imitation of new movements connected with parts of the body not visible to the child. (1) The model touched his forehead with his forefinger; child J. (11 months old at that time) watched this with interest and then put her right forefinger on her left eye, then over her eyebrow; in the next moment she rubbed the left side of her forehead with the back of her hand, and then touched her ear and came back towards her eye. After three further trials, when the child was already one year and 16 days old, she finally seemed to discover her forehead. When the model touched his forehead, she first rubbed her eye, then touched her hair, then brought her hand down and finally put her finger on her forehead. When this demonstration was repeated on the next days, the child at once succeeded in imitating this movement and even found the approximate spot on her forehead corresponding to the spot touched by the model. (2) The model touched his nose with his forefinger; child L. (one year and 19 days old at that time) tried to imitate him at once; she raised her forefinger and moved it in the direction of her mouth; she touched her lips, then moved her hand above the mouth and, finally, she found her nose.

Stage VI: Deferred Imitation

This stage refers to the cases when the imitation does not occur immediately after the demonstration by the model, but some time later and not necessarily in the presence of the model. According to Piaget's explanation, "imitation is no longer dependent on the actual action, and the child becomes capable of imitating internally a series of models in the form of images or suggestions of actions" (p. 62). Here are a few observations of this kind of imitation.

(1) Child J. (one year and 4 months old at the time) had a visit from a small boy (one year and 6 months old) who screamed when he tried to get out of a playpen and pushed it backwards, stamping his feet. Child J. never witnessed such a scene before. The next day she herself screamed in her playpen and tried to move it, stamping her foot lightly several times in succession. This was a clear imitation of the boy's behavior.

(2) After another visit from the same boy, child J. showed imitation of another of his behaviors. Namely, "she was standing up and drew herself up with her head and shoulders thrown back and laughed loudly" (p. 63) just as the boy did during his visit.

(3) Child J. also began to reproduce some words "not at the time when they were uttered, but in similar situations, and without having previously imitated them" (p. 63).

The cases of imitation quoted above are only a few examples chosen from his book (Piaget 1962). These examples show the general methods as well as main results of his work. The method consisted of close personal observations made by the author on the subjects (mostly his own children) from their birth through the first twenty months of life. The results showed that the sporadic ability to imitate is already present in the first few months of life and it develops together with the development of the child.

Piaget's observations were confirmed by the results of studies of other investigators. For instance, a review of research on imitation of live and televised models by young children by McCall et al. (1977) showed that infants younger than one year of age were already able to imitate simple motor behaviors with objects; the imitation of gestures was more common than the imitation of vocalization. However, the imitative ability depended on the child's level of mental development. Children below two years of age were not able to imitate sequences of behavior or to delay performance even after a short time after modeling. Children below three years of age imitated live models more than televised models; no such differences were found in older children.

Extensive studies on immediate and deferred imitation in 14 and 24 month-old infants were conducted by Meltzoff (1985). The first of these studies was conducted on 60 normal 2-year-old infants, with equal number of males and females. Each child was randomly assigned to an "immediate" group of 30 subjects, or a "deferred" group of another 30 subjects. The infants of each group were then randomly assigned to one of three test conditions: the baseline control, the activity control, and the imitation test, each of them conducted on 10 infants. All tests were performed individually for each infant, in an unfurnished room; the experimenter and the parent holding the infant faced each other across a small black table.

In the immediate imitation test, the experimenter (demonstrator) brought a toy from below the table, pulled it apart with a definite circular movement, and then he reassembled it. This act, which lasted 20 seconds each time, was repeated three times. After the demonstration the toy was lowered below the table edge and immediately brought back and placed in a spot 17 centimeters away, directly in front of the infant. The 20-second response period started from the moment when the infant touched the toy. In the deferred imitation test, the only difference in the procedure was that there was 24-hour delay interval between the demonstration and the response time.

The control baseline condition consisted of the same three presentations of the toy but without pulling it apart. In the activity control condition the toy was shown to the infant and was moved in a circle (as in the imitation test) but was not pulled apart. Each of these control procedures was repeated three times.

The results of this study indicated that imitation indeed took place. The statistical analysis of the data showed that immediate imitation of the model occurred in 80 percent, and deferred imitation in 70 percent of the 10 subjects of each group. On the other hand, only 20 percent of both immediate imitation control group and deferred imitation control group performed similar response.

In further research, Meltzoff (1988) tried to answer the question as to whether or not deferred imitation can occur in infants below one year of age. The methods used for this research were generally the same as those described for the previous study (Meltzoff 1985), except that not one but three toys were used. They were: (1) an L-shaped wooden construction in which manipulation consisted in pushing the vertical extension so that it lay flat on top of the base ("hinge holding"); (2) a small black box with a black button pushing of which produced a beep ("button pushing"); and (3) a small orange plastic egg filled with metal nuts which rattled when the egg was shaken ("egg rattling").

The first of these studies ("immediate imitation") was conducted on 60 normal 9-month-old infants, with equal numbers of males and females. The three toys with actions related to each of them were demonstrated three times, one at a time, for 20 second period, to each of 24 infants of the imitation group. The control groups (total 36 infants) were shown the three toys, but without the model's action related to each toy. The statistical analysis of the data showed that the number of correct responses of immediate imitation was significantly higher in the imitation group than in the control groups of infants.

The second study ("deferred imitation") was conducted on another group of 60 normal 9-month-old infants, equal in number of males and females. General procedure was the same as that used in the first study, except that there was a 24-hour delay in the response test. Specifically, the three objects together with the corresponding actions (hinge folding, button pushing, and egg rattling) were demonstrated, one at a time, and then the infants were sent home and were scheduled to return for the test 24 hour later. For the imitation test, each infant was shown again the three toys, as during the first visit.

The results of the deferred imitation test showed that the 9-month-old infants in the imitation group produced significantly more correct imitation responses than the control groups of infants. In other words, the 9-month-old infants appeared being capable of producing the correct manipulations with the objects shown to them 24 hours earlier by the model. This study suggests, therefore, that the capability of deferred imitative behavior (related to new objects) already exists in infants at below one year of age, that is, much sooner than was found in the earlier studies.

3

Imitation in Children Over
Three Years of Age

This chapter describes some cases of imitative behavior related to the previous experience and the present conditions of the imitation.

A study by Bandura et al. (1963) was designed to investigate the imitation of aggressive behavior and the effect of punishment of aggression on the imitative response. The study was conducted on 80 3 to 5-year-old children, 40 boys and 40 girls, from a nursery school. The children were assigned to one of four groups: aggressive model-rewarded, aggressive model-punished, a control group which was shown highly expressive but nonaggressive models, and a second control group which had no exposure to models. The models and their actions were shown to the first three groups on a film.

The aggressive model-rewarded group was shown an aggressive boy, Rocky, who encounters Johnny playing with his highly attractive collection of toys. When Rocky is refused when he asks to play with Johnny's toys, he becomes aggressive and strikes Johnny several times with a rubber ball and later with a baton; then he lassoes Johnny with a hoola hoop and pulls him forcefully to a far corner of the room. At the end of the film, Johnny is shown as sitting dejectedly in the corner while Rocky is playing with Johnny's toys, serving himself 7-Up and cookies. A commentator proclaims Rocky the victor.

The second group of children was shown the same film but with a different conclusion. That is, Rocky is severely punished by spanking for his aggression. At the end of the film Rocky sits cowering in the corner while Johnny takes his toys and leaves. The commentator announces that Rocky has been punished.

The first control group of children was shown another film in which two males are engaged in vigorous but nonaggressive play. They throw a ball to each other by bouncing it off the wall; they also use guns as a low hurdle in a jump game, they dance with a plastic doll, and roll and twirl a baton, and so on. The second control group of children was not shown any model film.

All the children were then tested for aggressive behavior. Each of them was placed in a room containing a variety of toys, such as a baton, two large Bobo dolls, hoola hoop, plastic farm animals, a lasso, dart guns, and so on, which could reproduce the aggressive behavior of the models seen before. Each child spent twenty minutes in the test room.

The analysis of the results showed that the behavior of children during the test was significantly influenced by the story in the model film seen before. Namely, imitative aggression was the highest in the first group which had been shown the film where aggression was rewarded. But it was only half that in the second group which had been shown a film where aggression was punished. The control groups which were shown nonaggressive film or no film at all showed more nonaggressive than aggressive responses.

The topic of another study by Bandura et al. (1966) was the effect of symbolization and incentive sets on observational learning. The subjects of the study were 6 to 8-year-old children, 36 boys and 36 girls, drawn from two elementary schools. The children were assigned to one of three groups: facilitative symbolization, passive observation, and competing symbolization conditions. Half of the subjects of each group were assigned to the incentive-set condition while another half to the nonincentive-set condition. The children of the facilitative symbolization group were instructed to verbalize every action of the model they would see in the movie.

The passive observation group was instructed only to pay close attention to the movie, and the competing symbolization group was instructed to count "1 and a 2 and a 3 and a 4 and a 5" repeatedly while they closely watched the movie. In addition, the incentive-set group was told that they would get candy for each item they reproduced correctly. The nonincentive-set group was only told that they would return to their classroom immediately after the movie.

The movie consisted of a four-minute color film in which an adult male model exhibited several unusual patterns of behavior such as building a tower in a specific way and placing a juice jar on the top, then firing a dart gun at the plastic container. Other performances of the model included several actions using a large Bobo doll, bean bags, hoola hoop, and so on.

In the following test, conducted in another room by another experimenter (who was not informed about the group assignment of the subjects), the children were asked to repeat all the actions shown by the model. It was found that the children of the concurrent verbalization group reproduced more matching responses than the passive observer group; the lowest number of correct responses was performed by the competing symbolization group. No difference in performance was found between the incentive-set group and nonincentive-set group. This study, therefore, demonstrated that the introduction of a competing verbal action during a viewing of the model's performance disturbed the delayed imitative process.

4

Role of Imitation in Cognitive Development and in the Therapy of Phobias

Learning by copying parents' or other adults' actions by a child is well known to occur both at school and at home. A child is constantly under the influence of the adults' behavior which he unknowingly or knowingly imitates. Undoubtedly, the imitation of others plays a critical role in the development of walking, eating, speaking, and other functions during the early stages of the child's life. Later, in school, the child is constantly shown what to do and how to write, read, and count. The teachers serve as models to be imitated. Direct modeling is used by instructors in a variety of sports, including sport games, swimming, or gymnastics. Imitation methods are also used in dancing lessons and "etiquette" behavior learning.

Imitative behavior is also common in adulthood. The styles introduced by fashion designers are widely imitated by people, and even choice of a specific restaurant or movie theater often depends on the current fashion.

The research on learning by imitation in humans includes a large number of publications. As the problem of learning by imitation is not the main subject of this book, we will describe only a few examples of this research.

The observations of Wishart (1986) were conducted on 24 6-month-old and 24 12-month-old infants, 12 males and 12 females in each group. Half of the subjects in each group were an experimental group and the other half a control group. Two cups were placed in front of the infant and a toy was hidden under one of them, and the infant was given two minutes to retrieve it. Then the toy was again placed under the same cup. After the infant was successful in finding the toy, a trial with hiding the toy under the other cup was performed. Each infant was given four trials, two starting on the left and two starting on the right cup. A new object was always used for each new set of trials. In a further stage, the toy was placed under one of the cups and then the cups were transposed, so that the toy was invisibly displaced from left to right, or vice versa. All groups were then retested on the same task one week later. Retesting of the experimental groups was preceded by a correct demonstration of the task by the infant's preschool sibling; the control groups were not given any model demonstration prior to reassessment. It was found that the performance showed a significant improvement after modeling, and that sibling modeling could facilitate cognitive development of the infant.

Tryon and Phillips (1986) studied a case of observational learning in three autistic boys, 4.0 to 4.7 years of age, with deficits in ability to learn simple tasks and games through observation of their peers, social interaction with peers, and appropriate play with toys. Six other boys at the approximately same age, were selected as peer models.

First, in order to establish the baseline design, each subject was escorted into a play area in which 10 unfamiliar toys and a peer model were present. The model played with each toy for 45 seconds at a time, with a 15-second interval after each play. In order to maintain the child's attention, the experimenter praised the child and occasionally offered him a small edible treat. If the observing subject did not show any imitative behavior during baseline, he was provided with training based on peer modeling.

During the training the subject observed a second peer model who demonstrated the targeted toy-playing twice. The peer model received praise or approved edibles after each correct response that he demonstrated for the observer. Then the subject was given the same toy and instructions. If he did not play with the toy appropriately on the first trial, the modeled demonstration was repeated and the child was retested. But, when the subject played correctly with the toy, he was given four additional trials to test for criterion of 5 consecutive correct responses. The reinforcement was given as before. Temporal maintenance was assessed at one and three weeks later. It was found that after the training, the number of imitative play intervals was increased in all three subjects, from an average of 2 during baseline to an average of 23 during the final maintenance test.

Carroll and Bandura (1985) studied the role of timing of visual monitoring and motor rehearsal in learning by observation. The study was conducted on 30 male and 30 female undergraduate students. Each subject watched a video monitor showing a male model performing a complex action pattern which engaged 9 different components related to the arm, wrist, and paddle, and took 33 seconds to execute. Another video monitor was used to play back the action pattern. The subjects were instructed to watch the television monitor at all times. The correct grip for holding the paddle handle was then demonstrated by the experimenter who also moved the subject's arm to correspond to the correct starting position twice. The subjects observed the modeled action pattern 12 times and were told to reproduce it after every two presentations. The statistical analysis showed that the reproduction accuracy was the highest with concurrent monitoring and motor rehearsal, and the lowest with no monitoring and no rehearsal. The data also suggested that in all trials the motor rehearsal produced significantly superior reproductions than no rehearsal.

In a further study Carroll and Bandura (1987) examined the role of visual guidance in observational learning. The subjects were 20 male and 20 female undergraduate students. The modeled action and paddle device were the same as that used in the former study (Carroll and Bandura 1985). The subjects were randomly assigned in equal number to one of four conditions. Depending on their assignment, they had to: (1) match a modeled action pattern concurrently with the model display, or (2) do that after the modeled display, (3) visually monitor their performance, or (4) not monitor it during the model's absence. It was found that both concurrent matching of modeled actions and visual monitoring significantly increased the level of observational learning. It was also found that the subjects maintained their performance skill even when the modeled and visual-monitoring guidance were withdrawn.

Several observations showed that modeling appeared helpful in reducing various fears in children, such as dental and animal phobias, fear of darkness, fear of water, and other phobias (see a review by Graciano et al. 1979). One such case was described by Osborn (1986) who studied the effects of modeling and desensitization on childhood warm water phobia. The study was conducted on a 6-year-old boy who grew fearful of warm water after he was scalded when he was 4 years old. He strongly resisted bathing and would eventually cry at the mere sight of steam. Play therapy and guided imagery did not reduce this phobia.

Then a modeling method was applied. The adoptive parents of the boy served as models. In the morning the mother prepared water 5 cm. deep at 70°F and entered it with the boy in her arms. Gradually, she lowered and withdrew the boy until he could remain seated at the bottom and bathe or play for three minutes. Then the water temperature was gradually increased by 1° every 3 minutes. A similar session was repeated in the evening when the adoptive father of the child prepared 5 cm. deep water at the maximum temperature used in the morning. The father entered the water first and gradually allowed the boy to contact and sink into the water. The depth of water was increased by 1.3 cm. every 3 minutes thereafter. The final phases were conducted in a sauna tub in order to attain greater depth of water. As a result of this procedure, the boy achieved bathing in 84 cm. deep water at 100°F for 3 minutes or more. A 6-month follow-up check showed that the boy bathed without visible fear and no longer reacted fearfully to steam.

Ost (1989) described the one-session treatment of twenty female patients suffering from various kinds of "specific phobias" such as injection (7 cases), spider (7 cases), rat (2 cases), cat (2 cases), bird (1 case) and dog (1 case). All these phobias were of long duration, from 6 to 39 years, in various patients. The method consisted of combination of two treatments: "exposure in vivo" and modeling. In the exposure in vivo, the patient was encouraged to approach the phobic stimulus and continue to expose herself to it until the anxiety decreased. In the modeling treatment, the therapist demonstrated to the patient his own interaction with the phobic object. The patient first touched the therapist while he had been touching the phobia object, and then, gradually approximated to full contact with the object. Then the patient acted more and more on her own, with the help of the therapist's instructions and during his presence in the room. The mean duration of the therapy session was 2.1 hours. The patients with animal phobias needed significantly longer sessions (2.4 hours on average) than the patients with injection phobias (1.6 hours).

The results of this combined therapy appeared to be very successful. The spider phobia decreased from 23.9 to 7.9 points. The injection phobia decreased to the mean 1.9 (of the scale ranging from 1 = no fear to 5 = maximum fear) which corresponds to 78 percent reduction. The other animal phobias also showed great improvement, on average a 65 percent reduction. The patients were then periodically tested for the maintenance of the therapy effects for a mean follow-up period of 4 years. Two of these patients (10 percent) showed only some improvement, 5 (25 percent) were much improved and 13 (65 percent) were completely recovered at the follow-up. Altogether 90 percent of the patients showed a clinically significant improvement; this improvement was maintained at the follow-up checking after an average of 4 years.

The cases of learning by imitation in humans described above indicated that imitative behavior plays an important role in the development of the organism by facilitating the acquisition of knowledge about the environment and behavioral adjustments to it. Some observations on learning by imitation in animals will be described further in this book.

5

Complex Cases of Imitation

As mentioned earlier, imitation occurs not only in the strictly defined cases of specific movements, but also in complex situations where the whole event is being copied. Imitation may occur, for instance, in arranging a specific ceremony, or party, in a way similar to that seen before at a friend's home. Also, a decision to send a child to a particular college may be dictated by observation of neighbors who have just made such decisions for their children, and so on. There are also some drastic cases, monitored by the media. Some time ago, we learned that an airplane highjacker took the ransom and, in addition, demanded two parachutes to jump from the plane; not long after this event two other highjackers used the same method in their crime. Although it is not certain that this was a case of imitation, it is not excluded that it could be one.

Imitation of Suicide

Some cases of complex imitation are especially dramatic. One of them is imitation of suicide. Let us describe at least a few examples of such unfortunate imitative behavior.

Mass media often bring news about suicide. A few years ago, the Japanese media reported the suicide of an eminent and popular female singer who because of some personal distress jumped from a high hotel window. A few days later several other young women committed suicide using the same way of killing themselves; in their suicide notes they indicated that they had to do so in support their dead idol's decision. In another case which took place in the United States, a young couple came to a small town and committed suicide by means of carbon monoxide exhaust from their car; they left a note asking to be buried together in one grave. A few days later, another young couple came to the same town and requested a similar funeral after their planned suicide which they wanted to be a copy of that committed by the first couple. However, the inhabitants of the town refused to cooperate and sent the couple back home.

These stories, however, were not sufficiently documented and it is not certain whether the suicide attempts of the followers were a result of the imitation of previous suicide victims, or an independent event. Recently, the problem of suicide, especially as it occurs among adolescents, has attracted special attention of physicians and psychologists.

Over two hundred years ago, Johann Wolfgang von Goethe wrote his famous book *The Sorrows of Young Werther* (1774) in which he describes with great artistry and emotionality a great unhappy love and the resulting suicide of a young man. The book became very popular in Europe but was blamed for leading some young people to suicide. Goethe himself wrote, "as I felt eased...at having transformed reality into poetry, my friends...thought they must transform poetry into reality, imitate a novel like this in real life, and, in any case, shoot themselves...this book which had done me so much good, was condemned as being highly dangerous" (cited in Eisenberg 1986). The sale of Goethe's book was bannned by authorities in Leipzig, and its publication in Copenhagen was forbidden. In Milan city officials bought and destroyed all copies for fear of imitative suicides (Eisenberg 1986). Nevertheless, the fame of the book endured and its influence on the decision to commit suicide is known as the "Werther effect" (the term proposed by Phillips 1974).

The Effects of Televised News Stories about Suicide

Some contemporary reports suggest that television news stories about suicide may be responsible for an increase in the suicide rate, especially among young people. Bollen and Phillips (1982) analyzed detailed daily mortality statistics in the United States (1972-76), based on the data from the National Center for Health Statistics. They found that the number of suicides increases by 7 percent during the week following suicide stories shown on two or three of the major television networks' evening news programs; this increase appeared to be significant when compared with the control period. The authors also tried to answer the question as to how long the effect of the suicide stories may last. Using the replication technique and the F-ratio, they found that the suicide stories have their effect for the first ten days after being shown.

In another study, Phillips and Cartstensen (1986) examined the effects of 38 nationally televised stories about suicide, broadcasted between 1973 and 1979, on the number of suicides among American teenagers. They found that the number of teenagers' suicides occurring within seven days after the broadcasts was significantly greater (1666) than the number expected (1555); that is, it was higher by 110 after these stories. They also found that the more networks carried the news, the greater was the increase in the number of suicides after the broadcasts. In the same study, the authors examined six alternative explanations of the increase in suicide rate, namely (1) prior conditions; (2) precipitation, not causation; (3) misclassification; (4) effect of grief; (5) seasonal effects; and (6) statistical artifacts. After analyzing the data and each possible alternative explanation, the authors concluded that none of these explanations seemed to fit the available data. According to them, the best available explanation of the results is that the televised stories on suicide triggered additional suicides by teenagers, and that the evidence is consistent with the supposition that "imitation and modeling are involved."

Gould and Shaffer (1986) studied the effect of four fictional stories about suicide that were broadcast on television. The first of these movies showed two male students who made a suicide pact; one of these students committed suicide by driving his car off a cliff. The second movie showed the male high school student who committed suicide after experiencing various interpersonal problems. The third movie presented a teenager who tried to stop his father's suicide. The fourth film was a story of a teenage boy and girl who jointly committed suicide. These films were part of the suicide prevention and educational program. Scripts of the films were distributed to be used in schools across the country. The telephone numbers of local suicide hotlines were displayed during and after the film. At the end of the film, the actress who played the suicidal boy's mother, appeared and told the viewers about books on the problem of teenage suicide.

The result of the above described program was quite the opposite of the expectations of the organizers. Namely, no decrease in suicide was observed and instead, a significant increase in the number of both completed and attempted suicides by adolescents during the two weeks after the broadcast, as compared with the suicide rate during the two weeks preceding the broadcast. The highest increase (from 17 before the movies to 33 after the movies, as estimated from the diagram) of both attempted and completed suicides took place after the first film. The high increase of the completed suicides also occurred after the third and fourth movies. The authors concluded that "the results are consistent with the hypothesis that some teenage suicides are imitative."

Motor Vehicle or Railroad Fatalities as Suicide Cases

Phillips (1979) examined the number of fatal motor vehicle accidents which took place between 1966 and 1973, during the period one week after a publicized front-page suicide story of each of 23 well-known persons and compared it with the number of fatalities that occurred during the period one week before the story. It appeared that the number of motor fatalities was significantly increased, on the average by 9.12 percent, in the week after the suicide story. The highest increase, 31 percent, was observed three days after the publicized suicide stories (figure 5.1). Further analysis of the motor vehicle deaths by Phillips (1979) revealed that the more publicity given to the suicide story, the greater the increase

FIGURE 5.1

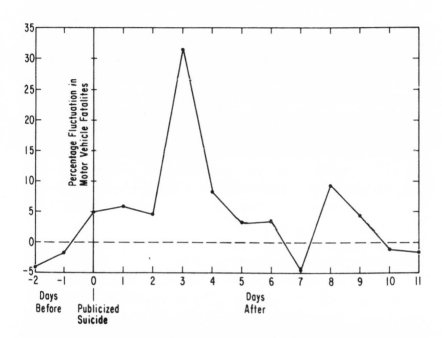

A Graph Showing Daily Fluctuation in Motor Vehicle Accident Fatalities for a Two-Week Period Before, During, and After Publicized Suicides.

Source: Reproduced from D.P. Phillips "Suicide, motor vehicle fatalities, and the mass media: Evidence toward a theory of suggestion," *American Journal of Sociology* 84: 1150–1174, 1979, by courtesy of the author and permission from the editor of the *American Journal of Sociology.*

in motor vehicle deaths. He also found that suicide stories about young persons are followed by an increase in single-vehicle accidents involving young drivers, and suicide stories about older persons tend to be followed by crashes involving older persons. Moreover, stories about murder and suicide tend to be followed by more multiple-vehicle accidents with passenger deaths, while stories about single-car accidents are followed by a higher increase in single-car accidents rather than in multiple-vehicle crashes. This indicates that each of these suicide stories could trigger a number of suicidal crashes exactly similar to the publicized suicide type.

Phillips (1979) also tried to find out where the fatal accidents after suicide stories take place. An examination of the corresponding data showed that after a suicide story motor vehicle fatalities were abnormally frequent mainly in the area where the suicide story was publicized.

A further study by Bollen and Phillips (1981) was undertaken as a replication of the earlier research (Phillips 1979). The previously obtained numerical data were analyzed again with the use of two different statistical methods. This replication of the former research fully confirmed the previous conclusion, namely, the authors found that motor vehicle fatalities in Detroit increased by 35 percent to 40 percent on the third day after a publicized suicide story, and that Phillips' previous finding was not an artifact or a result of improper technique of analysis.

Another case of the imitative effect of a television broadcast showing the railway suicide of a 19-year-old male student was described by Schmidtke and Hafner (1988). During a period of 70 days after the broadcast, the number of railway suicides increased sharply, mostly among 15 to 29-year-old males, but not among men older than 40 and women over 30. The control data from the same period during each of 5 years (1976–1980) preceding the year of the broadcast (1981) and 3 years (1981–1984) following it, show significantly lower numbers of suicides for 15 to 29-year-old males.

Other Cases of Imitation in Suicide

A "contagious" suicidal attempt among hospitalized adolescents was described by Kaminer (1986). A 17-year-old girl was admitted to the psychiatric in-patient unit after she unsuccessfully attempted to kill herself by pushing a long pin into her heart zone (pericardium). During the first week of her stay in the hospital this girl (given a diagnosis of major depression with melancholia) attempted suicide again, using the same method; this second attempt again was unsuccessful. However, within twenty-four hours, another girl in the hospital used the exact same method in an attempt to commit suicide but was also unsuccessful.

Another case reported by Ostroff et al. (1985), involved an adolescent boy and girl who had taken an overdose of drugs in an suicide attempt modeled after a televised story of the double suicide of an adolescent couple. Over the next two weeks, 12 more adolescents with drug overdoses were admitted to psychiatric or pediatric units. They did not know each other. The number of these patients (14) was considerably higher than the usual number of suicides per month (2 to 4). This finding indicated that the increase of suicide attempts was a result of the television story about the suicide.

Controversial Comments

Several authors expressed some doubts concerning the conclusions about the influence of television movies on suicide. Wasserman (1984) reexamined the findings of Phillips (1974) that the suicides of famous persons were followed by an increase in suicidal deaths. Extending Phillips data set to 1977, Wasserman (1984) found that, indeed, the stories on prominent suicides could result in a subsequent rise in the suicide rate. However, according to Wasserman, this rise could be related to the business cycle, occurring during a downturn in the economy in periods of unemployment and war. He concludes that "no significant linkage can be found between the national suicide rate and stories on prominent suicides on the front page of the *New York Times*."

Williams et al. (1987) also express some doubts about the interpretation of an increase in the number of suicides after a televised story on a drug overdose, as observed at Hackney Hospital and St. Bartholomew's Hospital in London. They found that the data from both hospitals did not provide sufficient evidence for the conclusion about imitative behavior.

Halasz (1987) criticizing the explanation of the clinical data as the "Werther effect," suggests that it is necessary to "recognize individual vulnerability in normal developmental phases, as occurs in childhood and adolescence... or dramatic changes in personal degrees of freedom."

Ostroff et al. (1985), who described a possible case of imitation in the drug overdose suicide attempts (as cited above), are still not certain about their conclusion. They comment that "all of these adolescents had underlying psychological problems, and several had attempted suicide before. Nevertheless, the movie may have given them a model for action that previously was absent." Likewise, Gould et al. (1988) strongly suggest that there is a need for further careful research in this area.

The instances of imitation in suicide described above show that such imitation does occur. However, it does not take place in all suicide cases. After a TV broadcast showing a story about suicide there was an increase in the number of suicides but certainly not a mass suicide of all viewers of the broadcast; this means that only some viewers were prone to imitate the story, while the others were not. Young persons were more likely to be influenced by the broadcast story about suicide than older persons. The increase in suicide was higher in locations close to the broadcast station than in other locations. Also, the post-broadcast suicide cases included persons suffering from stress and depression. All these examples of selectivity in the imitation of suicide suggest that a special critical factor plays an important role in imitative suicide behavior. This factor may be a preexisting condition such as depression, as was indicated by some of the above-mentioned investigators.

6

Effect of Companions on Feeding

Observations made on animals have provided much information about imitative behavior. Most of these data are related to feeding behavior. Therefore, this chapter will concentrate mostly on the results of studies on imitation in feeding in some birds and mammals. As before, several typical examples of research on animals, although not a full review, will be described.

A brief review of selected papers on vocal imitation in birds will be added at the end of this chapter.

Observations on Newly Hatched Chicks

An inventive method of studying the effect of the mother on feeding in chicks was used by Turner (1964). One-day-old chicks were singly placed in experimental compartment and were offered orange and green grain scattered on the floor of the compartment. Nearby, behind a wire netting, a model of a hen cut out of cardboard was placed to serve as an artificial mother (figure 6.1). By a means of a lever, the model was made to perform pecking movements at the grain. When the grain at which the model was repeatedly pecking was orange, the chicks pecked twice as much at the orange grain than at the green grain. When the grain at which the model was pecking was green, the chicks pecked significantly more at the green than at the orange grain. Here, the influence of the model on the choice of specifically colored grains was clearly visible.

FIGURE 6.1

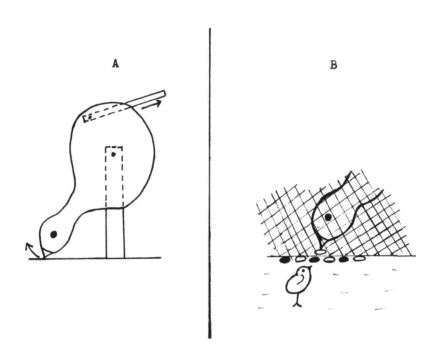

Imitation of the Model Hen by a Young Chick in Selection of Grain.

Sketch A shows the mechanism used for producing pecking-like movements of the mechanical hen with a manually operated lever (arrow). Sketch B shows the mechanical hen pecking at the grain behind a wiremesh partition, and a chick observing the hen.

Source: Reproduced from figure 15 ("The mechanical hen") and figure 16 ("The mechanical hen pecked at a green colored grain and the chick had the choice of green or orange grains") from E.R.A. Turner "Social feeding in birds," *Behaviour* 24: 1–46, 1964, by permission from the publisher of *Behaviour.*

More recently, the findings of Turner (1964) were confirmed by Suboski and associates (Bartashunas and Suboski 1984, Suboski, 1984, 1987, 1989, Suboski and Bartashunas 1984). In their studies on the pecking of hatchling chicks, they used an angular stimulus ("arrow") that made vertical "pecking" movements at a distinctive food-like object. Visual exposure to the pecking arrow enhanced ability to elicit from the chicks pecking at similar small distinctive objects. It was also observed that the chicks pecked significantly more at stimuli matching in color the objects in the display that were attached to the pecking arrow or to the floor beneath the arrow, than at other stimuli. There was greater enhancement of responding to a stimulus color during both up and down movements of the arrow than during other movements. In this study the role of the arrow was the same as the role of the model hen in the imitative behavior of chicks in the study of Turner (1964).

Here, we should also quote the earlier observations of Tolman (1964) who found that pecking in chicks can be elicited by sounds similar to pecking but produced by an object different than a companion's beak. Namely, when the investigator tapped on the floor with a pencil, the chick approached and started to peck at the pencil; the chick could even be led away from the food by tapping some distance from the food. Attempts to peck at the pencil were observed even in the case when a glass divider was placed between the tapping pencil and the chick.

We will discuss the possible mechanism of this phenomenon later in the theoretical chapter of this book.

The Effect of Presence of the Mother or Other Adults on the Development of Independent Feeding in the Young

The observations on food intake in young animals showed that independent feeding develops under the influence of parents or other adults of the colony. In a study by Galef and Clark (1971), adult rats were offered a choice of two diets. One of these diets was contaminated with toxic lithium chloride. After the consumption of the contaminated diet and the resulting sickness, the adult rats started to avoid that diet. They continued to avoid it even when it was no longer contaminated with the poison (cf. Rzoska 1953).

When the rat pups in their early weaning period were allowed to accompany the adults at feeding time, they ate only the diet that was being consumed by the adult rats and avoided the diet that was being avoided by the adults.

In another study, Galef and Clark (1972) observed that rat pups given a choice of foods in the absence of the adult rats of the colony did not start to eat solid food until they reached a mean age of 25.5 days. But when the rat pups were offered the same choice of food in the presence of the adult rats, they began to accept solid food much earlier, at the mean age of 19.9 days.

The same researchers also found that the choice of food by the rat pups was not always dependent on the diet which they were given during nursing. When the adults were present during feeding, the pups showed preference for the feeding site where the adult rats had been eating, even when the diet eaten by the adults was different than that consumed by their mother during the nursing period. This observation suggests that the presence of the adults was a stronger stimulus in food preferences than the cues related to the early feeding with the mother (Galef and Clark 1972).

In further research, Galef (1977) examined the factors influencing the selection of feeding sites by rat pups. He found that the members of adult pairs of rats did not attract one another to a feeding site. He observed, however, that the juvenile rats, in most cases, chose the feeding site where an adult rat or a juvenile sibling had been present. This observation suggests that a tendency to seek the companions and eventually to imitate them occurs mostly in young animals and disappears later in life.

The Effect of the Companion(s) on the Amount of Food Consumption

Katz and Revesz (1921) reported that chickens that had been fed in isolation until they were satiated and would no longer accept food, started to eat again as soon as they saw other birds eating (cited after Thorpe 1963). Harlow (1932) found that rats ate

FIGURE 6.2

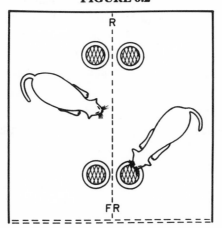

A sketch of the experimental situation used for studies on the effect of companion on food consumption in cats. Each cat can choose from eating from a container located close to the companion (visible behind the wire-mesh partition) or from a container located far from the companion.

FIGURE 6.3

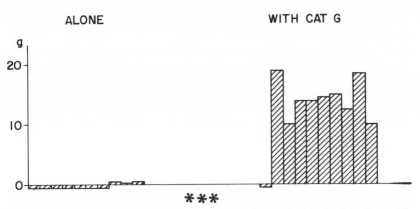

The effect of companion on consumption of new, less palatable food (pellets with cereal). The columns show the amount of food consumption in solitude (left) and in the presence of the companion eating food (right), in 10 consecutive sessions of each series. Columns below 0 show no consumption. Asterisks at the bottom indicate the statistical significance of the data.

more when they were fed in groups than when they were fed in isolation. Similar results of studies on the effect of companions on food intake in rats were obtained by Soulairac and Soulairac (1954). Tolman (1964) observed that a pair of chicks offered food first started to peck at it for a while, independently from one another. Then, a pause in pecking followed. When one of these chicks started to peck again, the other chick approached and started to peck at its companion's beak and then to peck at food. Such behavior resulted in an increase in the amount of food intake in chicks reared in pairs. In another study Tolman (1968) observed that the effect of the companion on pecking depends on the behavior of the companion. When the companion was inactive (e.g., sleeping) the amount of pecking of the observing chicks was significantly lower than when the companion was active. However, no significant differences in pecking were found between the cases when the companion's activity was related to feeding and the cases when this activity was different than feeding. More recently, Johanson and Hall (1981) also reported that rat pups increased their milk intake in the presence of their siblings.

The effect of a companion on the amount of consumption was also studied on adult cats (Wyrwicka 1990). One of these studies was conducted on five cats in a compartment divided with a wire mesh fence into two even parts, A and B. Each cat occupied a separate half of the compartment. The cats could see and smell the companion but could not touch each other. Each cat participated in several series of 10 sessions each. In some experimental series one cat was placed in part A and the other in part B (series with companion). In another series one cat was placed in one part (A or B) while the other part remained empty (series without companion). During a 10-minute session cats, deprived of food for 4 hours, were allowed to eat canned tuna from any of two cups located at the dividing fence, 30 cm. apart from each other, in each part of the compartment (figure 6.2). In series with companion, each cat could choose to eat either from the cup located just across from the cup from which the other cat had been eating, or from the more distant cup. The amount of consumption from each cup was determined by weighing the cup with tuna before and after the session. A statistical analysis of the data (using the t-test) showed that the mean consumption in 7 of 10 series with companion was significantly higher than the mean consumption in the series without companion. Figure 6.3 shows the data obtained from a cat who practically refused to eat probably less palatable food (meat pellets with cereal) when eating alone but accepted a large amount of the same food in the presence of a companion. No significant differences were found in the remaining 3 series. Neither were there significant differences between consumption close to and far from the companion.

However, in one pair of cats, the presence of a companion appeared to be an inhibitory factor, resulting in a lower food consumption. We will return to this case in chapter 10 of this book.

In order to obtain more experimental data concerning the eventual tendency to seek companionship, another 10-session series was conducted with 4 of these cats. The cats were tested in a Y maze arranged on the floor of the laboratory room. Each cat played a role of either an "actor" or a "companion" in various sessions. The actor cat was placed in a "waiting" cage and allowed to observe two arms of the maze for 10 seconds. At the end of each arm a cup filled with tuna was placed at the wire mesh fence. At one arm, left or right in various sessions, behind the fence, there was another cat (the companion) who had been eating from the cup. The actor cat was then released from the waiting cage and allowed either to go to eat from a cup placed across from the companion or to go to eat in isolation from a cup placed at another arm of the maze. No significant differences were found between the numbers of times the cats preferred eating with a companion or alone. This shows that in this case seeking companionship was not an important factor in choosing the site of feeding. This confirmed the observations of Galef (1977) who found that the "members of adult pairs of rats did not attract one another to a feeding site." Thus, these studies on adult cats suggest that the increase of consumption in the presence of a companion was independent of the tendency to seek companionship.

Another study dealt with the effect of the presence of companion on consumption of milk with an addition of alcohol (Wyrwicka and Long 1983). The research was conducted on two adult cats ("drinkers") who voluntarily drank small amount of 10 percent ethanol solution in milk; three other cats ("nondrinkers") served as companions to the drinkers. The experimental sessions were conducted in the same double compartment described above, except that only the frontal part of the compartment was used.

The cats were introduced into the experimental compartment either singly or in pairs. Each pair consisted either of two drinkers or one drinker and one nondrinker. Each cat of the pair was placed in one part of the compartment. Each drinker was offered 10 percent ethanol solution in milk, while each nondrinker was given plain milk. A series of sessions with a drinker was followed by a series with a nondrinker or with no companion. There were a total of 13 series of sessions for each drinker. A statistical analysis of the data showed that, in most series, the mean amount of consumption of ethanol solution was significantly higher in the presence of a companion (either drinker or nondrinker) than in its absence.

A Case of Interspecies Imitation in Feeding

A unique and fascinating observation on social influences on food preferences was made in a zoo in Japan. Dr. Yojiro Kawamura from Osaka University (now at Koshien University) reported to this writer that herbivora such as deer or giraffes sometimes accept fish when they have been raised in the same open area with fish-eating birds such as flamingos or pelicans (figure 6.4). This fish-eating behavior in herbivora was observed mostly in young animals, and in the wintertime. This fact suggests that food preferences in animals of one species may be influenced by animals of other species when they are living to-

FIGURE 6.4

Deer Eating Fish in a Zoo in Japan.

Source: Reproduced from an original photography by courtesy of Prof. Yojiro Kawamura.

gether in the same area. It is possible that, in this case, the whole heterogeneous group of deer, flamingoes, and pelicans can be considered one colony. Therefore, the members of the colony, regardless of whether or not they belong to the same or different species, may influence the food preferences in the young, similarly as adult rats influence feeding in rat pups in their homospecies colony. Similar cases of induced food preferences can be found in pets living with people.

Addendum: Vocal Imitation in Birds

The capacity of parrots for vocal imitation is well known; they can imitate human speech sometimes with great perfection. Some birds, such as starling (*Sturnus vulgaris*), the mocking bird (*Mimus polyglottus*) and the marshwarbler (*Acrocephalus palustris*) include notes and phrases of songs of other species into their own song; the grackle (*Gracula religiosa*) can perfectly imitate not only human speech but also a great variety of various noises (Thorpe and North 1965). Studies on singing abilities of birds showed that their songs (whose usual function is to serve as territorial proclamation) are individually acquired by imitating the parents or other members of the community. For example, chaffinches (*Fringilla coelebs*) reared in isolation were able to produce only a very primitive type of song which could not be divided into phrases. However, when they were reared in groups with other young chaffinches, they produced more elaborate songs, divided into phrases. Their song patterns were usually similar to those of other members of the group, but clearly different from the songs of other similar groups. This observation suggests that learning by imitation could take place within each group of young chaffinches reared in isolation from the adult birds (Thorpe 1961, Hinde 1969).

When chaffinches reared in isolation were exposed to tape recording of chaffinch song during their first autumn and winter, they produced almost normal songs later. However, this does not mean that they can learn any song they hear. They can only incorporate limited foreign song patterns in their full song. Once a full song has developed, no further song patterns are learned (Hinde 1969).

Perhaps it would be interesting to quote a few observations on the use of the learned song by some birds, described by Thorpe and North (1965). Namely, it seems that most of the outstanding vocal imitators are found among tropical and subtropical species. The function of the song "is apparently as a social signal for maintaining pair and family bonds and is part of the sexual display rather than a territorial one." In 1903 Waite reported that two captive Australian magpies (*Gymnorhina tibican*) learned to sing a fifteen-note melody played to them on the flute, as antiphonal duet. When the younger bird died, the survivor resumed the performance of the whole song, which it had never been heard to produce during the years when it had a companion (quoted by Thorpe and North 1965).

More data on this topic can be found in publications of Guillaume (1971, orig. 1926), Thorpe (1961), Thorpe and North (1965), Hinde (1969) and Davis (1973).

7

Role of the Mother in Initiation of Eating New Food by Weanling Kittens

This chapter describes the research which was conducted in our laboratory in the past years (Wyrwicka 1978, 1981, 1988; Wyrwicka and Long 1980). The issue under investigation was whether or not the weanling kittens would accept new food independently of the presence or the absence of their mother.

The observations were conducted on four mothers and their 19 kittens. The kittens were 30 to 39 days of age at the start of the research. The new food was canned tuna for six 32-day-old kittens of mother M and four 30-day-old kittens of mother S (the tuna group). Another food was cream of wheat cooked with an addition of vegetable oil for five 39-day-old kittens of mother E and four 36-day-old kittens of mother O (the cereal group).

Each litter was divided into two subgroups. One subgroup of kittens was placed in an experimental compartment and was offered the new food in the presence of the mother who had just been eating that food. The other subgroup of kittens of the same litter was, at the same time, placed in another experimental compartment, similar to the other one, and offered the same food in the absence of the mother. In each compartment the food was offered on a white paper plate, 20 cm. in diameter. This secured an easy access of the kittens to the food. The animals were allowed to eat ad libitum during the session. The results of the study were the following (figure 7.1).

FIGURE 7.1

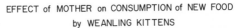

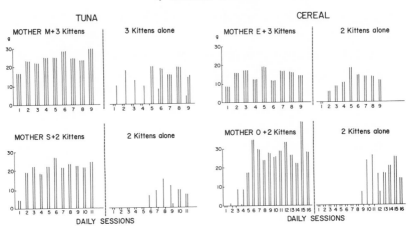

Amount of consumption of a new food (tuna or cereal) by weanling kittens in the presence and in the absence of the mother (Wyrwicka and Long 1980).

The Tuna Group

Kittens with the Mother

Observations were made on five kittens (three kittens of mother M and two kittens of mother S). When the kittens of each litter were placed in the experimental compartment together with their mother for the first time, they showed a brief orienting reaction (up to 1 minute), and then they started to explore the environment. For the first 2 to 3 minutes they did not seem to pay any attention to the mother who was consuming tuna from the plate located at the center of the compartment. Then they started approaching the mother. A few seconds later they began to lick the tuna. One kitten chewed the rim of the paper plate for a while before it started to lick some tuna. It was observed that at first the kittens tried to eat tuna from the exact same spot on the plate where their mother had been eating. Only after a few such attempts they started to eat from another spot on the plate; however, they continued to keep close to the mother. Usually, their brief episodes of eating alternated with longer episodes of play and searching around.

During further sessions, the orienting and investigatory reactions were still observed at the beginning of each session but they grew shorter and shorter in the following sessions. Finally, beginning with the sixth or seventh session, the kittens started to eat tuna immediately after having been introduced into the experimental compartment. They usually ate continuously for 3–5 minutes, then stopped eating and went off to play; from time to time, however, they returned to eat a little more.

Kittens Alone

Each of the two remaining groups of siblings (three kittens of mother M and two kittens of mother S) was placed in the separate experimental compartment apart from their mother (who, in the meantime, remained in another experimental compartment together with the other group of her kittens, as described above). The plate with tuna was placed in the center of the compartment's floor. The kittens first showed a strong orienting reaction. They simply "froze" as if frightened, and remained motionless for about three minutes. Only then did they start to investigate the environment and eventually, to play. They seemed to completely ignore the plate with tuna. Toward the end of the session only one kitten (that of mother M) approached the plate and started to eat tuna. The same kitten ate tuna also during the following days, usually in the first few minutes of the session. The other kittens of both litters, however, continued ignoring the plate with tuna for at least four days. Starting with the third day, however, they were occasionally observed approaching the plate with tuna and playing with it by pushing it with paws. It was observed that playing with the plate immediately preceded the start of consumption of tuna. Namely, during the play with the plate and chewing its rim, some tuna juice entered the kitten's mouth and the kitten swallowed it. In the next moment, the kitten would start to lick the tuna on the plate and then actually to eat it. In the sessions that followed, regular consumption of tuna took place.

However, the initiation of eating tuna in two of these kittens required more sessions. They apparently avoided any contact with the plate for five days (the kitten of mother M) or even for eight days (the kitten of mother S). While their siblings were already eating tuna, these two kittens were only observing them from a distance. Finally, on the sixth day, the kitten of mother M took a small piece of tuna from the floor after a piece had fallen from the plate (during the consumption by the sibling) and ate it. Since then, this kitten ate tuna with its siblings at each following session. The kitten of mother S was even more resistant to accepting tuna. While its sibling had been eating, this kitten moved restlessly around the plate, but did not try to eat. Finally, it started to consume tuna on the ninth day.

These experiments showed a clearcut difference between the behavior of kittens with the mother versus alone with siblings. While the kittens with the mother initiated eating the new food, that is, tuna, on the first session, most kittens without the mother started eating tuna only after several consecutive sessions.

The Cereal Group

Kittens with the Mother

The experimental design was the same as that with the tuna group, except that instead of tuna, the kittens were offered cream of wheat (cooked in water and cooled before the session) with a small addition of broth flavor. In order to induce the mother to eat this food, mother E and mother O received a rewarding mild electrical stimulation in the lateral hypothalamus every two or three seconds while eating cereal during the session. This method is described in detail in the next chapter.

Mother E was accompanied by three of her kittens and mother O by two of her kittens during each session. The behavior of the kittens was quite similar to that of the kittens offered tuna, as described above. However, there were some differences in the time of initiation of eating cereal. While the kittens of mother E, after a brief (30 sec) orienting reaction, started to eat cereal during the first session with the mother, the kittens of mother O initiated eating cereal only after a few sessions. These differences could be caused by the tendency of these kittens to play during the whole session. Their initiation of eating cereal was also preceded by playing with the plate (on which the portion of cereal was placed). However, once they initiated eating cereal, they ate it with the mother at each of the following sessions.

Kittens Alone

The observations were made on the remaining siblings, two kittens of mother E and two kittens of mother O. Similarly to the behavior of kittens alone in the tuna group, the kittens alone in the cereal group showed first a strong orienting reaction (lasting about two minutes) and then started to investigate the compartment and later to play. They completely ignored the plate with the cereal. The kittens of mother E began their eating cereal already during the second session and ate cereal in increasing amounts in each following session. The kittens of mother O, however, seemed to ignore the plate with cereal for the first several sessions. One of these kittens began eating cereal only during the fifteenth session, and the other during the eighteenth session. Afterwards both kittens ate cereal during each following session. The amount of their consumption however, was lower than the amount of consumption of their siblings eating cereal with the mother.

8

Imitation of the Mother's Improper Food Selection by Weanling Kittens

The preceding chapter showed that the initiation of eating new food was strongly influenced by the presence of the mother who had just been consuming that food. When the mother was present and eating the new food, her weanling kittens started to consume it during the first or the next session with the mother. But when the same new food had been offered to the siblings of these kittens in the absence of the mother, they did not accept it until after a number (up to eighteen) of sessions during which the new food was offered. A question arose whether the influence of the mother includes the case when the mother has been consuming food improper for the species, that is, the kind of food which is normally refused by the adult cats.

In order to answer this question, first the kind of the improper food had to chosen. For cats, one such food was bananas. Offering it to adult cats in our laboratory evoked a reaction of turning the head away from the banana slices, and other signs of disgust, including escape from the observation area by some of the tested cats. However, a previous study showed that it was possible to induce eating bananas in cats by the use of the self-stimulation method (Wyrwicka 1974), as described below.

The Method of Inducing Eating an Unusual Food in cats

In an adult cat, under pentobarbital anesthesia, two monopolar electrodes (made from stainless steel wire, about 0.3 mm. in di-

ameter, insulated except for 1 mm. at the tip) were implanted bi-
laterally in the lateral hypothalamus. A jeweler's screw was fixed
in the frontal bone (over the frontal sinus) to serve later as a reference electrode. Several weeks after the implantation, when the cat
fully recovered from the operation, a test for self-stimulation was
performed. The cat was placed in an experimental compartment
equipped with a lever connected to an electric stimulator. Each
time the cat approached, sniffed or touched the lever, a brief (0.3
second) train of impulses (2 to 3 volts, 100 impulses per second, 1
millisecond duration per impulse) was given through the tip of the
hypothalamic electrodes. The stimulation was given first manually by the experimenter until the animal started to press a lever
for more stimulation (which evidently was desirable to the animal). After several sessions with pressing a lever for stimulation
the experimental procedure changed. The lever was removed from
the compartment and, instead, two kinds of foods: meat pellets
and banana slices were offered, each on a separate plate. The cat,
deprived of food for 4 hours preceding the test, was moderately
hungry, and usually started to eat meat pellets immediately after
entering the experimental compartment. During brief intervals in
eating pellets, the cat occasionally approached the plate with bananas and sniffed it. Each time this happened, the previously ef-
fective hypothalamic stimulation was given. In a few minutes, the
cats stopped eating meat pellets and concentrated on sniffing, touching, and licking banana slices, each time obtaining the rewarding
stimulation. After a few following sessions, the cat started to eat
bananas as soon as it entered the experimental compartment, completely ignoring meat pellets.

The experimental technique described above was used in the
project designed to answer the question as to whether the weanling kittens would be influenced by the mother in eating the improper food (bananas). In female cats, in the early period of
pregnancy, the operation of implanting two electrodes in the hypothalamus was performed, as described above. After implantation, each mother-to-be was housed in a large compartment and

fed ad libitum with a variety of foods such as meat pellets, skim milk fortified with protein, and canned tuna or chicken.

About one month after delivery, each mother was taken for the self-stimulation test (as described above). The mothers that learned to press the lever for the rewarding hypothalamic stimulation were used in further experimentation. Each mother, always deprived of food 4 hours prior to the session, was offered meat pellets and banana slices (each food on a separate plate) and the desirable hypothalamic stimulation was given only for eating bananas (and never for eating meat pellets). After a few daily sessions, the mother usually started to eat banana slices while ignoring meat pellets as soon as she entered the experimental compartment. Eating was usually continuous, with occasional brief intervals. The hypothalamic stimulation was given approximately every 2 seconds during the usual 10 minute session.

Eating Bananas by Kittens in the Presence of the Mother

When the mother's behavior stabilized, the kittens, one or two at a time, were introduced to the compartment together with the mother. The age of the kittens varied between four and ten weeks. The kittens accompanied the mother during the session three to five times a week, for at least four weeks. The kittens as well as their mother were food deprived for 4 hours before the session.

The kittens introduced with the mother to the experimental compartment were allowed to behave as they pleased. It was observed that first they showed an orienting reaction by remaining motionless close to the entrance door for the first 30 to 60 seconds. Then they started to investigate the environment. After several minutes they started to play with various small objects (a screw head or a small hole in the wall, or the mother's tail, etc.). They did not seem to pay any attention to the mother who had been eating bananas, and they also ignored meat pellets (the food known to them from their home compartment). The same behavior was observed during the following sessions. However, their orienting-investiga-

FIGURE 8.1

A kitten eating slices of bananas with the mother. The mother is induced to eat bananas by the hypothalamic stimulation reward.

tory reaction became shorter and shorter with each following session, until it completely disappeared.

After a few sessions with the mother, the kittens started to pay some attention to their mother. They began approaching the mother more and more often; finally, they started to lick and then actually eat the banana slices from the mother's container (see figure 8.1). Initially they were keeping very close to the mother and ate only from the same spot of the plate where their mother had been eating. In the following sessions, however, they also ate from another spot on the plate, and later, even from another plate located close to the mother's plate or from the floor

TABLE 8.1

A comparison of consumption of bananas by mother C ("poor eater")
and mother L ("good eater") and their kittens before and after weaning.
(Means for Five 10-Minute Sessions)

Mother Alone			Kittens Alone (after training with mother)		
No.	g*	No.	9–16 weeks of age	17–23 weeks of age	25–27 weeks of age[†]
C	3.4±0.4	5C	1.2±0.5	no data	no data
		6C	3.3±1.0	no data	no data
		7C	2.8±0.5	10.5±1.6["]	10.5
		8C	2.6±0.9	14.0["§]	35.0[§]
		9C	1.7±0.6	no data	no data
L	39.0±3.4	10L	18.0±1.5	36.1±7.4	35.0[§]
		11L	11.5±2.5	13.6±4.8	19.2
		12L	6.5±1.4	6.0±2.7	17.5
		13L	3.1±0.7	2.1±1.0	10.5
		14L	3.9±0.8	5.4±1.3	0.0

*Mean consumption of banana (in grams per session ± SE).
†Single 24-hour test in the home compartment.
"Mean for five 24-hour tests in the home compartment.
§All amount of banana given to the kitten was consumed.

when a banana slice had accidentally fallen from the plate. Usually, the kitten (or kittens, if a pair of them accompanied the mother during the session) started to eat with the mother's plate immediately after the entering the experimental compartment and continued eating during the first 2 to 3 minutes. Then, they started to play around, only occasionally returning to the mother's plate to eat bananas with her for some seconds.

Eating Bananas by Kittens in the Absence of the Mother after Weaning

When the kittens reached the age of three months, they were separated from the mother. They were, however, still taken for a session twice a week. Each kitten (or a pair of them), were always deprived of food for 4 hours prior to the session, as before. It was placed in the experimental compartment and both kinds of food, banana slices and meat pellets, were offered on separate but identical plates. It was found that most kittens which had previously eaten bananas with the mother continued to do so in the absence of the mother. Immediately after entering the experimental compartment they approached the plate with banana and started to eat. After eating almost continuously for 1 to 3 minutes, they walked away to play; however, they occasionally returned to eat for a short while. They completely ignored meat pellets.

The amount of consumption differed from kitten to kitten and from litter to litter. It was observed that the kittens whose mother usually ate only a small amount of banana during the session, ate less banana than the kittens whose mother ate much (table 8.1). It was found, however, that these differences in the amount of banana consumption were present only during the early postweaning period and were no longer observed when the kittens were tested at six months of age. At that time, most of these kittens ate even more bananas than at the early age.

Similar results were obtained with other unusual foods, such as potatoes and flavorless jellied agar, when they were offered to the kittens in the presence of the mother who had been eating those foods. Experiments with each of these foods will be briefly described below.

Eating Potatoes by the Mother and the Kittens

This series of experiments was conducted on only one mother and her four kittens. A premeasured portion of plain mashed potatoes (without any additions) was offered on a paper plate placed at the center of the experimental compartment's floor. A portion of meat pellets was also offered on another identical paper plate, located nearby the plate with potatoes. First, the mother was trained to eat the potatoes for the hypothalamic stimulation reward in a series of sessions. Then a pair of 5-week-old kittens, was introduced to the compartment together with the mother. The kittens first showed an orienting reaction (manifested mostly by a motionless posture) which after some seconds was replaced with investigatory behavior. They did not pay attention to the mother's eating potatoes. In the following sessions, the orienting-investigatory behavior became shorter and shorter. Finally, one of these kittens started to eat potatoes with the mother and since then it ate potatoes during each following session. The other kitten, however, refused eating potatoes. It was once observed playing with a piece of potato which fell from the plate, but did not eat it. During the session this kitten was several times seen suckling on the mother (while the mother was eating potatoes).

In further series of sessions, another pair of kittens, who were already nine weeks of age, was introduced to the compartment with the mother. These kittens were no longer nursed by the mother but they were still living with the mother in the same home compartment. They were food deprived for 4 hours prior to the session. After the entrance to the experimental compartment, they showed only a brief orienting reaction, and then they approached the mother who was eating potatoes. A few moments later, they joined the mother in eating potatoes. They did not seem to pay any attention to meat pellets which were offered on another plate placed near to the plate with potatoes. Both kittens ate potatoes continuously for 4 to 5 minutes. One of them ate vigorously and sometimes vomited after a few minutes of continuous eating; in such case, this kitten immediately consumed the vomited potatoes and then returned to the plate to eat more potatoes. On an average, these kittens consumed about 10 grams of potatoes per kitten, during a 10 minute session. After several sessions with potatoes, these kittens were taken to be tested for food preferences in the absence of the mother. A portion of potatoes was offered on one plate and a portion of canned tuna on another plate. The kittens were already acquainted with tuna because this food (highly preferred by cats) was often offered to them in the home compartment in the presence of the mother. The kittens first rushed to eat potatoes and ate them vigorously. However, after about 10 seconds, they turned to eat tuna. After further 10 seconds of eating tuna, they returned to eat potatoes; such alternating between potatoes and tuna was repeated several times.

The experiments with eating potatoes by weanling kittens described above showed individual differences among the kittens in the acceptance of potatoes. These differences seemed to depend on the age of the kittens. One kitten tested at the age of 5 weeks started to eat potatoes with the mother only during the fifth session with the mother. The other kitten of the same age refused to eat potatoes at all; instead, this kitten was observed suckling on the mother during the session (as mentioned above). On the other hand, the other pair of kittens, tested at 9 weeks of age, started to eat potatoes already in the first session with the mother. This suggests that some maturity of kittens was needed for the acceptance of new, unusual food, even in the case when the mother had been consuming this food in their presence.

Eating Flavorless and Odorless Jellied Agar with the Mother

In order to check whether eating of unusual foods such as bananas or potatoes was a result of the imitation of the mother and not of an attractive taste or odor of the unusual food, a tasteless and odorless jellied agar was chosen to test the mother's influence on the kittens. This study was carried out on six mothers and their sixteen kittens. Each mother was first trained alone to eat agar for the hypothalamic stimulation reward. Then the kittens, singly or in pairs, were introduced to accompany the mother during the session. The pieces of cooked jellied agar were offered on a paper plate; nearby, an identical paper plate with meat pellets was placed.

During the first few sessions the kittens showed first an orienting reaction and then an investigatory reaction. Both these reactions were shorter and shorter in each following session. During the remaining time of the session, the kittens were playing. However, in a few further sessions, the kittens approached the mother and started to eat or at least to lick pieces of agar with the mother. The amount of consumption could only be estimated by subtracting the average consumption by the mother (measured previously when the mother was tested alone) from the total amount of consumption by the mother and the kittens. The mean consumption during each 10-minute session (of a total 12 sessions) was 2.2±0.8 grams per kitten, while the mean consumption by the mother was 7.1±3.1 grams.

The experiments with jellied agar showed that the weanling kittens will eat unusual food, such as jellied agar, without any flavor, if their mother has been eating that food.

It was interesting to check whether the addition of taste or smell to jellied agar will facilitate or inhibit the process of imitation of the mother by the kittens. Therefore, in separate series of sessions, various flavors (one at each series) were added to plain agar. These additives were the following: 0.001 percent quinine hydrochloride in distilled water, 1 percent saccharin solution in distilled water, and vegetable broth. Let us briefly describe these experiments, separately for each additive.

Agar with 0.001 Percent Quinine Hydrochloride

The test conducted prior to this series of sessions showed that the addition of 0.001 percent quinine hydrochloride to the diet produced an aversive reaction in cats. This experimental series was conducted on only one mother and her two kittens nine weeks of age. Before this experimental series, the mother was trained to eat plain jellied agar for the rewarding hypothalamic stimulation and ate it each time it was offered. Her kittens ate plain agar with the mother during approximately 50 percent of the sessions with the mother. Then a portion of agar sprinkled with the quinine solution on one plate, and a portion of plain agar on another identical plate (located near each other) were offered to the mother and the kittens. The rewarding hypothalamic stimulation was given to the mother only for eating the quinine-contaminated agar but not for eating plain agar. However, the mother refused to eat the agar with quinine during the first and the third session of the test, but consumed a little of it during the second session. The kittens licked the agar with quinine a few times during each session, but no actual eating was observed.

After three sessions with the mother, the kittens were offered the quinine-contaminated agar and plain agar, on separate plates, in the absence of the mother. This test was conducted during the 8 following sessions. It was observed that the kittens ate agar during only some of these sessions. One of these kittens ate agar with quinine during 6 out of 8 testing sessions, while the other kitten ate it during only two of the eight sessions. Usually, the kittens ate by licking alternately from the plate with quinine-contaminated agar and the plate with plain agar. The mean consumption of quinine-contaminated agar was estimated as 1 gram per kitten per session. The amount of consumption of plain agar was approximately the same.

Agar with 1 Percent Sodium Saccharin

Previous experiments showed that 0.2 percent saccharin solution in water was preferred to distilled water in adult cats (Wyrwicka and Clemente 1970). In this experimental series 1 percent commercial saccharin solution was used. The test with saccharin was performed on two other mothers (which were not used in the previous test with quinine) and their four kittens, two from each mother. In the preceding period each mother and her two kittens were given only plain agar, while the mother was given the hypothalamic stimulation reward for eating plain agar. The kittens ate agar in approximately 50 percent of the sessions. During the test with saccharin, a portion of jellied agar sprinkled with saccharin solution was offered on one plate, while the same amount of plain agar was offered on a separate, identical plate next to the other one. Each mother was rewarded by hypothalamic stimulation only for eating plain agar. The kittens ate alternately from both plates, although they spent more time eating plain agar with the mother. Some individual differences in consumption of agar were observed. One kitten of each mother ate agar with saccharin, alternating it with plain agar in most of a total of 11 sessions while the other two kittens ate agar with saccharin during only a few sessions.

Two of these kittens were then tested in the absense of the mother for eating agar with saccharin and plain agar. It was observed that one kitten ate saccharin-sweetened agar during each session, whereas the other kitten ate it only during one session. It seemed that the kittens did not differentiate between plain agar and the agar with saccharin, because they ate alternately from both plates. The amount of consumption of agar with saccharin was estimated as approximately equal to the consumption of plain agar (+2 grams per kitten).

Agar with Vegetable Broth

Observations were made on three mothers and their seven kittens. Each mother was previously trained to eat plain jellied agar for the hypothalamic stimulation reward. The kittens accompanied their mother during two or three sessions before the test with agar with vegetable broth. In the next session, pieces of agar on one plate were sprinkled with vegetable broth. In the next session, pieces of agar on one plate were sprinkled with vegetable broth (prepared from commercial vegetable broth powder) while pieces of plain jellied agar were offered on another, identical plate. It was observed that the mother who had been rewarded with the hypothalamic stimulation for eating plain agar, ate mostly this agar. However, she often interrupted eating and went to the other plate with broth-flavored agar, ate it for several seconds and then returned to eat plain agar. The kittens ate alternately plain agar and agar with broth; the amount of their consumption of broth-flavored agar was approximately equal to the amount of their consumption of plain agar. It was also observed that the kittens were more excited in the sessions with broth-flavored agar than in sessions with quinine- or saccharin-flavored agar. They ate broth-flavored agar during almost each session, while they were eating quinine or saccharin-flavored agar only in some sessions. The mothers also seemed to be more excited and they often ate alternately from the plate with broth-flavored agar and the plate with plain agar.

After weaning, four of these kittens (of two mothers), nine-week-old at that time, were tested for eating plain agar or broth-flavored agar, in the absence of the mother. It was observed that they ate mostly broth-flavored agar but also occasionally licked plain agar. The kittens ate agar during all sessions of the test, except for one kitten which refused to eat agar during one session. The amount of consumption of broth-flavored agar was 3.3±0.8 g per kitten per session (of total 5 sessions with two of these kittens, and 2 sessions with the other 2 kittens), but the consumption of plain agar was near zero.

Two of the kittens which had been trained with the mother in eating plain agar were tested again when they were already five months old. These kittens, after the completion of the experiments with the mother, were not given agar at all for the two month period preceding the test. Each kitten was placed separately in the experimental compartment and was offered pieces of plain agar on a familiar plate. It was found that each kitten immediately approached the dish with agar, sniffed it and started to lick it. They usually ate continuously for several seconds, and then went off to play, but returned to eat agar again. Several episodes of licking and actual eating agar were seen during intervals of play. It was found that one of the kittens consumed 2.8 grams and the other 1.1 grams of agar during the 5-minute test session. These results suggested that these kittens recognized agar as a type of food they had eaten earlier in life.

Test with Control Kittens

Four kittens, approximately four months old, which had never accompanied their mother in the session and had never been given agar or seen the mother eating it, were used for control. Each of these kittens were placed alone in the experimental compartment and offered pieces of jellied agar on a familiar dish. The 5-minute observation sessions were conducted daily, for 5 consecutive days. It was observed that during the first session, all of the kittens were excited. They approached the plate with agar and sniffed it. Two of these kittens also licked it a few times. Two other kittens, after initial approaching the agar, started to play around the compartment, but occasionally returned to sniff the agar for 1-2 seconds, without licking or eating it. Similar behavior of each of these kittens was observed during the second session. During three remaining sessions, the interest in agar seemed to disappear completely, and each kitten was only playing around the compartment without paying any attention to agar. This behavior was completely different from the behavior of the kittens which had been previously trained with the mother in eating agar, as described above.

The observations described in this chapter suggest that the kittens which accompanied the mother during the session with eating unusual foods (bananas, potatoes, jellied tasteless and odorless agar) learned to eat this food with her, and acquired a specific knowledge about this kind of food by imitating the mother in eating it. After weaning, in the absence of the mother, they recognized this food and ate it when it was offered. More information on this topic can be found in previous publications (Wyrwicka 1978, 1981, 1988).

The results described in this chapter can be taken into account as a warning related to food preferences in humans. The tendency of children to imitate adults includes the imitation of food preferences, even when the food may be harmful to health. The improper choice of food acquired by imitation of adults in early childhood may result in a damage to health later in life.

9

Learning by Observation

The famous experiments of Thorndike (1911) on animal behavior included an attempt to answer a question as to whether or not animals can learn a specific motor task solely by observing an animal demonstrator performing the task. Studies were conducted on chicks, cats, dogs and monkeys. The task for chicks was an escape from a pen, for cats it was pulling a string or climbing, for dogs jumping and biting a cord, and for monkeys manipulating a variety of devices. It appeared that none of these animals could learn the tasks solely by observing the demonstrator performing the required task.

However, later studies by other investigators provided positive findings concerning the question of animals learning by observation. Darby and Riopelle (1959) reported that monkeys significantly improved their discrimination of objects when allowed to observe another monkey receiving a reward for selecting a designated object. John et al. (1968) showed that cats who observed demonstrator cats jumping over a hurdle to avoid foot shock, or, in another experiment, demonstrator cats pressing a lever to get food, later learned these tasks with significantly fewer errors than cats trained without previously observing the demonstrators' performance. Chesler (1969) demonstrated that kittens allowed to observe their mother pressing a lever for food learned to perform this task sooner than kittens which did not observe the mother before training. In all these cases, motor acts were not acquired solely by observation; however, subsequent acquisition of the task was facilitated by observation.

More recently, Heyes and Dawson (1990) reported a case of learning by observation in rats. Experiments were conducted in a large chamber divided into two identical compartments by a wire-mesh partition. In the operant compartment, used for demonstrations and testing, an aluminium joystick was suspended from the middle of the ceiling. Pushing the joystick to the left or to the right (depending on the experimental design) by the demonstrator was rewarded with a food pellet. Half of the demonstrators were trained to push the joystick to the left, and another half to push it to the right. The observing rat was placed in the other compartment of the chamber.

The acquisition test showed that rats that had observed left-pushing made more left responses than rats which had observed right-pushing, and vice versa. In the reversal of a left-right discrimination, rats that had observed demonstrators pushing in the direction that had previously been rewarded took significantly longer to reach the criterion reversal and made more responses during extinction than rats that had observed the demonstrator's pushing in the opposite direction to that previously rewarded. All these results showed that rats can learn a specific bidirectional motor task through observation of the demonstrator rat performing this task.

Learning by Following the Leader

Several investigators studied a special case of animal behavior when the animal is following a companion to a goal. In experiments of Miller and Dollard (1941), rats were trained to follow a leader rat in a T-maze. The rats learned to imitate the leader in selecting the route in the maze but only when the correct response was rewarded with food each time, an observation confirmed by several other investigators. Bayroff and Lard (1944) reported a case of imitative learning by following behavior in white laboratory rats. The experimental sessions were conducted in a T-maze submerged in water; the rat had to follow a previously trained leader rat to select the correct arm of the maze. The correct response was rewarded by removing the rat from the water before removing the leader (to prevent the experimental rat from seeing the leader obtain the reward). Eight of 20 rats learned to follow the leader; some of these animals even learned to seize the tail of the leader to be guided. When the reward was obtaining food by hungry rats, all rats learned to follow the leader. This study represents a case of learning by imitation, but it is not an example of spontaneous imitative behavior.

Several other investigators also studied the problem of learning by following the leader. Solomon and Coles (1954) demonstrated that 97 percent of their experimental rats learned to imitate a leader in a T-maze for food. Church (1957) found that rats learned not only to follow the leader in a T-maze but also to respond independently (in the absence of the leader) to light cues that were previously used for guiding the leader.

Stimbert et al. (1966) demonstrated that rats also learned to follow a leader when experiments were conducted in an "open-field" situation with the use of an apparatus which increased the number of choice-point alternatives. In their experiments, rats were deprived of water 22 hours prior to the experimental session. The correct response of following the leader (which was, in advance, trained in finding the correct route in the maze) was rewarded by obtaining water. These studies not only confirmed the previous finding, but also extended it by demonstrating that rats can learn not only to follow other rats, but also to do so in a free environment where the number of alternative routes is twice that used in a T-maze.

More recently, Rabenold (1987) conducted a 5-year study of a partially marked population of black vultures (*Coragyps atractus*) on their feeding habit. This study showed that individual birds of this species can find food by following others from overnight roosting groups. Arrivals at the food site occurred primarily on the day of discovery and on the following day, but the incoming groups were larger on the second day. Adults arrived at the food source earlier that young adults and juveniles. Birds removed experimentally from the population sufficiently long to be ignorant of the location of food, followed others from the roost when reintroduced.

Simple wildlife observations provide more examples of following the leader or other members of the animal colony. Most commonly known is flying single file or formation flying birds colonies during seasonal migration, or the tendency in sheep to follow one another.

However, we may ask whether these behaviors are truly cases of imitation. The behavior of following the leader or another member of the colony does not appear to exactly fit the definition of imitative behavior, formulated at the beginning of this book.

Several investigators observed a tendency of newly hatched chicks, ducklings, or goslings to follow the parent or the first moving object (see a review by Hess 1973, pp. 67–70). This phenomenon was systematically studied by Lorenz (1935). He observed that newly hatched greylag geese followed their mother almost immediately after hatching. When, instead of the mother they confronted another moving object, for example, a human, they followed him in the following days. This first experience of the young birds also effected their behavior later in life. Lorenz (1935) called this phenomenon "imprinting," stressing that it did not result from learning.

Comparing this behavior with cases of imitation described in this book, we hypothesize that the mechanism responsible for the behavior of newly hatched birds which follow the first moving object can differ from the mechanisms of imitation (see theoretical considerations in chapter 11).

10

Inhibition of Imitative Behavior

Although imitative behavior seems to occur quite frequently, there are cases when imitation does not take place. The inhibition of imitation may be caused by various factors, such as antagonistic stimulation occurring at the time of the demonstration, former experience, or rules imposed by law.

Let us consider some examples of inhibitory behavior in imitation. As already mentioned in chapter 6, Galef (1977) found that social influences on feeding were observed mainly among young rats who most often chose the feeding site where an adult rat or a juvenile sibling had been present; the adult pairs of rats, on the other hand, did not attract one another to a feeding place. Similar observations were made on adult cats, which usually ate more in the presence of a companion, but did not show a preference for seeking companionship during eating, as mentioned in chapter 6 (Wyrwicka 1990).

An example of suppression of imitative behavior was observed in a study by Wyrwicka and Long (1983, described in chapter 6). Adult cats were offered milk with addition of 10 percent ethanol solution either in the presence of a companion which had been drinking milk (without ethanol solution), or in its absence. It was found that the cats drank significantly more milk with alcohol in the presence of a companion than in its absence. However, when the sessions were continued, the effect of the companion became irregular and finally disappeared. This suggested that some other factor, independent of the presence of the companion, could inhibit its former influence on the consumption. Possibly the odor of ethanol, which did not play any role earlier, started to be aversive when presented repetitively.

Another observation which may be taken into account as a case of inhibition, is the example of interspecies imitation in feeding, described earlier. According to the observer, Yojiro Kawamura, the imitative effect on eating fish by deer was observed mainly in young deer and mostly in winter. This means that older deer were resistant to accepting fish. Again, there must have been a specific factor which inhibited imitation of eating fish. This factor could be a previously acquired habit of eating other foods.

A question may also be asked as to why some weanling kittens did not imitate their mother in eating banana immediately, but rather only after several sessions. Evidently, in this case the presence of the mother was not a sufficient condition for producing imitative behavior. One factor responsible for the absence of imitation could be a strong tendency of these kittens to play instead of seeking food; this tendency was present in lesser or greater degree in all kittens. Another possible cause of the delay in imitation of the mother could be the behavior of the mother. When the mother was eating unusual food under the hypothalamic stimulation she did not pay as much attention to the kittens as during eating usual food (cf. Wyrwicka, 1981, pp. 42–43). This could result in a decrease of watching the mother by the kittens, leading to their engagement in play rather than in imitation of the mother.

The inhibition of imitation was also observed in a study on the effect of a companion on food consumption in adult cats. These experiments (using the technique described in chapter 6) were conducted in a large cage, divided with a wire-mesh into two sections. Each cat of the pair was placed singly in one section; the cats could see and sniff the companion but could not touch one another. As described in chapter 6, most cats ate more in the presence than in the absence of a companion. However, in one pair of cats, the presence of a companion appeared to be an inhibitory factor. In this case, each cat ate less in sessions with

FIGURE 10.1

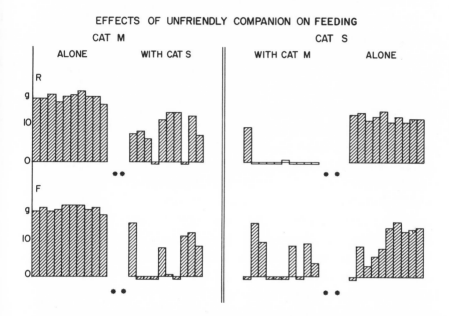

EFFECTS OF UNFRIENDLY COMPANION ON FEEDING

Effects of unfriendly companion on feeding.

A decrease of food consumption in the presence of an aggressive companion. The columns show the amount of consumption from both food containers (close or far from the companion) during each of 10 consecutive sessions. For other explanations see the legend for figure 6.3.

companion than in sessions without companion (figure 10.1). It was observed that both cats manifested aggressive behavior, hissing at each other. One of them would even occasionally beat at the divider with its paw as if it wanted to attack the other cat. In this case the imitative effect of the companion was inhibited by its aggressive behavior.

An experimental case of inhibition of imitative behavior was demonstrated in a study of Bandura et al. (1963) on imitation of aggressive behavior in young children, already described in chapter 3. In this study, the group of children who saw the film in which aggression was rewarded showed highly aggressive behavior when tested later. On the other hand, the aggressive behavior was significantly suppressed in the group of children who previously had seen the film in which aggression was severely punished. Here the punishment of aggressive behavior played a key role in inhibiting imitation of such behavior.

Another case of inhibition of imitation was demonstrated in a study on the effect of symbolization on observational learning in young children (Bandura et al. 1966, described in chapter 3). When children were instructed to count from 1 to 5 while viewing the model's performance, the number of their correct responses decreased and the delayed imitative process was clearly disturbed. Here the competing symbolization activity resulted in partial inhibition of imitation.

The existence of inhibition in imitation is supported by some clinical studies. Lhermite et al. (1986) found that 28 of the 29 patients with focal lesions of frontal lobes imitated the examiner's gestures such as bending the head and resting the chin on the hand, tapping the leg with the hand in time to various rhythms, kicking, crossing the legs, military salute, folding a sheet of paper and putting it in an envelope, chewing paper, writing, drawing, and others. The patients imitated these gestures even though they were not instructed to do so. When asked why they did it, they answered that they thought that they had to imitate the examiner. After being told not to imitate, most patients continued imitating. No such behavior was observed in normal subjects or in patients with different brain lesions.

These observations suggest that the frontal lobes may play a role in the inhibition of imitative behavior. When this brain area was damaged due to surgery, the imitative behavior clearly increased. Similar disinhibition of the previously acquired (by training) inhibitory behavior was observed in earlier studies of the effects of prefrontal lobectomy on motor conditioned reflexes in dogs

(Brutkowski et al. 1956). This suggests that the inhibitory function of frontal lobes may be more general.

Let us hope that further systematic research on the problem of inhibition of imitation, so important in social life, will determine the exact conditions in which such inhibition takes place.

11

Summary and Theoretical
Considerations

Summary

Imitative responses in human neonates

Newborn babies, as young as 0.7 to 71 hours of age, were able to imitate facial gestures of the adult demonstrator, such as mouth opening and tongue protrusion.

Neonates at a mean age of 36 hours were able to imitate the adult's model facial expressions such as a happy smile, sad protrusion of lips or opening of the mouth suggesting surprise.

Observations on infants at 6, 10, and 14 weeks of age showed that tongue protrusion can be produced not only by the demonstration by the adult model but also by a black pen and white ball moving in the front of the infant.

Matching behavior to the tongue protrusion model declined by 12 weeks of age in these infants.

Hand opening and closing by the human model could produce the same response in a 14-week-old infant. This response could also be elicited by a dangling ring.

Imitative Behavior in 6 to 16 months old children

Babies between six and eight months of age could imitate sounds such as "pfs," "bva," "hha" made by the adult.

Babies between eight and nine months of age were able to imitate the movements of the demonstrator even when their own movements were not visible to them.

Older infants, 1-year-old or a few months over one year, were able to imitate more complex movements such as drawing on a sheet of paper. Children of this age were also able to reproduce some movements one day after the demonstration (deferred imitation).

Imitative Behavior in Over 3-Year-Old Children

The aggressive behavior of 3 to 5-year-old children was increased after a projection of a film in which aggression was rewarded; on the other hand, aggressive behavior was decreased after showing the film in which aggression was punished.

The introduction of a competing verbal action during viewing the model's performance disturbed the delayed (deferred) imitative process.

Learning by Observation

A number of experimental studies on both children and adults showed that observation of the performance of others facilitated the subsequent acquisition of the behavior shown by the model.

Imitation of Suicide in Humans

Statistical analyses of suicide attempts and completed suicides showed that a television broadcast showing a story about suicide often led to a rise in the number of suicides. The highest increase was found in the first few days after the broadcast. The suicide method used in each case was the same as the method shown in the film, such as railroad fatality, a suicidal car "accident," or drug overdose.

The increase in the number of the post-broadcast suicide cases was higher in the areas located close to the television station than in other areas, and it occurred more often in the young than in the old persons. Also, the post-broadcast increase in suicide included persons suffering from stress and depression.

Imitation of Feeding Behavior in Animals

Newly hatched chicks imitated the model mother or a model "arrow" in pecking significantly more at the grain of color selected by the model mother or the arrow than at the grain of the other color.

Pecking could also be evoked in chicks by sounds of tapping with pencil on the floor.

Chickens fed *ad libitum* in isolation started to eat again when they saw other birds eating.

Rat pups imitated the adult rats in choosing the kind of food being consumed by the adult rats, and avoiding the food which had been avoided by the adults.

The Effect of the Companion on Food Consumption

Rat pups ate more when fed in groups than when fed in isolation.

Adult cats ate more in the presence of a companion than in its absence.

Young deer and giraffes in a zoo sometimes accepted fish when they lived in the same area with flamingos.

The Effect of the Mother on the Initiation of
Eating New Food by Weanling Kittens

Weanling kittens offered a new food (canned tuna, food usually preferred by adult cats) started to eat it in the first or the second session when this food had been offered in the presence of the mother. However, all but one of their siblings who were offered the same food in the absence of the mother, initiated eating only after four to nine (in various kittens) daily sessions. When the new food was cereal, and was offered in the absence of the mother, one kitten initiated eating it only after 14 sessions and the other after 17 consecutive sessions.

The Effect of the Mother Eating an Improper Food on Consumption of that Food by Weanling Kittens

When the mother was artificially induced (by rewarding hypothalamic stimulation) to eat food unusual for her species, such as bananas or tasteless and odorless jellied agar, the weanling kittens also ate that food, although usually only after a few sessions with the mother. These kittens accepted this unusual food also after weaning, in the absence of the mother.

Inhibition of Imitative Behavior

Imitative behavior in adults, both human and animal, was observed only in some individuals but not in all of them. The imitative responses were suppressed after showing to subjects (children at 3 to 5 years of age) a film where aggressive behavior was punished.

The food selection responses acquired by weanling kittens through the imitation of the mother were seen only in some but not in all kittens when they were tested several months later.

The imitative effect of the companion's drinking milk on consumption of the mixture of milk with alcohol in cats was decreased after several repetitions of the test.

Theoretical Considerations

Let us now discuss a theoretical basis explaining the facts described above. The phenomenon of imitation was analyzed and discussed by a number of scientists at the beginning of the twentieth century. At that time, psychologists such as Baldwin (1906) and McDougall (1908) expressed a view that imitation is basically an inborn ability, an instinct. According to the explanation of the imitation of simple movements in infants at an early age by McDougall (1908), "we have to assume the existence of a very simple perceptual disposition to having this specific motor tendency, and, since we cannot suppose such a disposition to have been acquired at this age we are compelled to suppose it to be innately organized. Such an innate disposition would be an extremely rudimentary instinct" (p. 91).

Later investigators, such as Guillaume (1971, orig. 1926), concluded that imitation is a complex phenomenon in which instinct is modified by associative factors. The systematic research on imitative behavior of infants and children at various age, undertaken by Piaget (1962, orig. 1951, described in chapter 2 of this book) pointed to the influence of developmental factors on imitative behavior. Depending on age, the child is able to imitate more and more complex acts. Piaget's observations led him to conclude that imitation "is controlled by intelligence as a whole. To put it more exactly, it is reintegrated in intelligence, since imitation always has been the continuation of the accommodation of the schemas of intelligence, and it is when the progress of this accommodating mechanism is in equilibrium with that of mental assimilation that the interaction of these two processes replaces imitation in the general framework of intelligent activity" (Piaget 1962, p. 78). In his further considerations he concludes that imitation "fits into the general framework of the sensory-motor adaptations which characterize the construction of intelligence." He continues by saying the following: "Sensory-motor intelligence is therefore always both accommodation of the old schema to the new object, and assimilation of the new object to the old schema"(p. 84). However, Piaget emphasized that "although imitation always depends on intelligence, it is in no way identical with it" (Piaget 1962, p. 85).

A different approach to the problem of imitation was that of Thorpe (1963, orig. 1956). Basing himself mostly on the observations of animals, this investigator concentrated on social effects on imitative behavior. According to him, "much of so-called imitation amongst animals is either social facilitation or local enhance-

ment and probably does not involve any form of insight learning."
"Imitation as a whole…might be described as 'social learning',
and social facilitation can be described as 'contagious behaviour',
where the performance of a more or less instinctive pattern of
behaviour by one will tend to act as a releaser for the same
behaviour in another or in others, and so initiate the same lines of
action in the whole group" (Thorpe 1963, pp. 132–33). On the
other hand, Thorpe expresses a view that "imitation must be defined very carefully if the term is to have a useful meaning." Therefore, he introduces the term "true imitation." He explains that by
"true imitation is meant the copying a novel or otherwise improb-
able act or utterance, or some act for which there is clearly no
instinctive tendency." He doubts, however, that we can find any
certain examples of such behavior anywhere in animals below pri-
mates (except possibly in cats).

A Comment on the Terms "Social Facilitation"
and "True Imitation"

The terms "social facilitation" and "true imitation" introduced
by Thorpe (1963) have been accepted by a number of authors.
However, because the research on imitative behavior is conducted
by both psychologists and neuroscientists, it seems that it would
be useful to try to "translate" these psychological-sociological
expressions into neuroscientific terms.

Let us then analyze the term "social facilitation." The word "social" indicates the presence of stimuli deriving from an individual
or a group of individuals usually belonging to the same species
who have been performing a specific behavioral task ("models").
The word "facilitation" suggests that models have an ability to
facilitate the performance of an observer who is trying to repeat
their activity. Several studies, however, showed that the presence
of models does not always facilitate the performance; on the con-
trary, it may disturb the activity of the observer. For instance, when
the model was punished for aggressive behavior, the observers

strongly decreased their imitative aggressive activity thereafter (Bandura et al. 1963). In 24 studies involving 24,000 subjects (reviewed by Bond and Titus 1983), the presence of others had only a small effect (0.3 percent to 3 percent of the variance in the typical experiment); it was also found that the presence of others increased the speed of simple task performance but decreased the speed of complex task performance (Bond and Titus 1983). In a study on imitative feeding behavior in cats, it was found that when the companion, visible through a wire partition, was vigorously eating, the observer cat consumed more than when it was fed separately (Wyrwicka 1990). However, when the model cat behaved aggressively, hissing and beating at the partition with its paw, the consumption of the observer cat was significantly lower than when it was fed alone (Wyrwicka, unpublished data). Here, instead of facilitation, inhibition occurred.

These data suggest that the term "social facilitation" describes the general effect of the companion (or a group of companions) on the observer rather than strictly on the act of copying the behavior of the companion(s). It seems, therefore, that the term "social facilitation" does not properly describe imitative behavior. As to the term "true imitation," reserved by Thorpe (1963) to describe the cases of copying only an entirely new behavior, it does not seem that it refers to a completely separate neural mechanism other than the mechanism responsible for usual imitative reactions (such as tongue protrusion in neonates, or resumption of eating in satiated chicks when they see other chicks eating). In fact, the imitation of the "old" acts occurs very often; most cases of imitation involve previously performed behaviors. It seems, therefore, that the use of the term "imitation" for all cases, new and old, simple and complex, is actually most informative and comprehensive.

The Possible Brain Mechanisms Related to Imitative Behavior

Whereas in the past explanations of imitative behavior were based on ideas such as "instinct" and "insight," at present there is

FIGURE 11.1

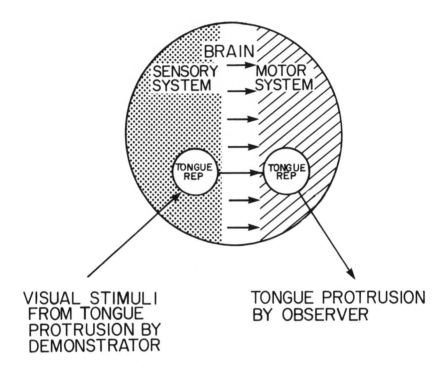

VISUAL STIMULI   TONGUE PROTRUSION
FROM TONGUE   BY OBSERVER
PROTRUSION BY
DEMONSTRATOR

A simplified sketch of unconditioned reflex arc for imitation of tongue protrusion in the young infant. Visual stimuli from the tongue protrusion by an adult activate the tongue representation (Tong Rep) in the brain sensory system of the infant; this leads, through sensorimotor connections (arrows), to an activation of the tongue representation in the infant's brain motor system, resulting in imitation of the tongue protrusion by the infant.

a tendency to look for the brain mechanisms which may be responsible for the phenomenon of imitation. An example of this tendency is a study by Meltzoff and Moore (1977). Based on their observation that infants at 12 to 21 days of age were able to imitate simple facial and manual gesture, they expressed a view that this early imitation must be based on the "neonate's capacity to represent visually and proprioceptively perceived information in a form common to both modalities." Their further observations of facial imitation in 40 newborns at 42 minutes to 71 hours of age led them to hypothesize that "both visual and motor transformations of the body can be represented in a common form and thus directly compared." "Infants could thereby relate proprioceptive/ motor information about their own unseen body movements to their representation of the visually perceived model and create the match required." They postulate, therefore, that "infants can recognize and use intermodal equivalences from birth onward" (Meltzoff and Moore 1983).

Brain Mechanisms Responsible for Simple Cases of Imitation

Speaking directly about the brain mechanism of imitation, let us first try to find the sequence of the events leading to imitation in human neonates. As shown in figure 11.1, the visual (or acoustic, in some cases) stimuli from the demonstrator's gesture (e.g., tongue protrusion) reach and activate the area corresponding to this gesture in the sensory system of the brain. The activation of this sensory area results in sending neural impulses to the corresponding area of the motor system of the brain, through the sensorimotor connections between these areas. (The existence of such connections between the sensory and motor system was demonstrated by electrophysiological research [Woolsey 1958], lesion studies [Stepien and Stepien 1959], and conditioning experiments [Dobrzecka and Wyrwicka 1960]; more recently, Miller and Vogt [1984], using the horseradish peroxidase labeling technique, demonstrated the existence of direct visual-motor cortical connections). As a result, the process of activation of the neonate's brain sensory and then motor areas corresponding to the demonstrator's gesture leads to the performance of the same gesture by the neonate.

Since imitative behavior occurs in the neonates at only one hour of age, it is reasonable to postulate that the imitative behavior of the neonate could not be acquired by experience but was a result

FIGURE 11.2

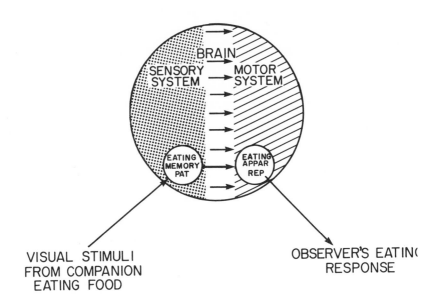

A simplified model of the mechanism responsible for imitation in eating behavior. Visual stimuli from the companion eating food activate a memory pattern of associations related to previous eating experience (Eating memory pat) in the brain sensory system of the satiated observer. This in turn activates, through sensorimotor connections (arrows), the representation of the eating apparatus in the observer's brain motor system (Eating Appar. Rep), leading the observer to resume eating.

of an inborn ability, due to the sensorimotor connections in the brain. A question arises as to how to classify this phenomenon. It seems that the simplest mechanism responsible for this imitative behavior is *the unconditioned reflex of imitation.* The use of the term "unconditioned reflex" implies that this reflex is inborn. This term corresponds to the term "instinct" which is considered an inborn ability, and is still used by some writers.

The finding that the neonate's imitative response of tongue protrusion may also be produced by a black pen or white ball moving in the front of the infant, and the response of hand opening and closing may also be elicited by a dangling ring (Jacobson 1979), just as pecking in newly hatched chicks can also be provoked by tapping with a pencil on the floor (Tolman 1964), support a view that imitation is based on an innate, unconditioned reflex. Evidently, the objects and activities used in the testing (pen, ball, dangling ring, tapping) possessed certain elements which were also present in the models' activities and were critical for the production of the imitative reflex in the neonates.

Brain Mechanism Responsible for Complex Cases of Imitation

Examples of complex cases of imitative behavior include resumption of eating in a satiated state only because the others are eating. Let us analyze this case. As was previously shown, the act of eating is a complex activity based on innate reflexes (such as licking, mastication, or swallowing), which become adjusted to the type of food and the environmental conditions of consumption. Due to the repetitive occurrence of this act under the same environmental conditions, a specific pattern of associations between these conditions and the features of the food is established in the brain. In an example described earlier, a satiated chick that refused to eat any more when fed in isolation resumed eating when it saw other chicks eating (Katz and Revesz 1921). In this case, the sight of companions engaged in eating activated the pattern of associations in the brain sensory system, acquired during previous eating experiences; this resulted in the activation of the representation of the eating function in the brain motor system, through sensorimotor connections, leading to eating as schematically shown in figure 11.2. The same brain mechanism is responsible for the case when animals eat more when fed in groups than when fed in isolation.

The case of kittens imitating the mother in eating a new kind of food, even when this food is unusual for their species (e.g., bananas), represents a combination of the unconditioned reflex of imitation and the previous experience with eating. When the same situation, the presence of the cat mother eating the unusual food, occurs repeatedly, the initial imitative behavior is gradually replaced by an instrumental conditioned reflex. In such a case, the whole experimental situation including the mother and the food containers, becomes the complex conditioned stimulus, the eating of the unusual food becomes the conditioned response, and the satisfaction derived from eating becomes the rewarding reinforcement (Wyrwicka and Chase 1972). This enables the kittens to eat the same unusual food also in the absence of the mother.

This also refers to other cases of imitation when a new stimulus or action is demonstrated for the first time. In each of these cases, the initial imitative behavior is gradually replaced by the formation of an instrumental conditioned response (called "conditioned reflex type II" by Miller and Konorski 1928, Konorski 1967, and "operant behavior" by Skinner 1938). In other words, the process of learning by imitation takes place. This was convincingly demonstrated in studies of Piaget (1962).

There are some cases of imitation which lead to consequences dangerous to the organism, such as using drugs, smoking, or drinking alcohol to excess. But the most tragic is the imitation of suicide. Its brain mechanism is usually activated not by direct visual stimuli from the victim but by information from somebody else, or from mass media. Therefore, the imitative suicide belongs to the category of imitation by activation of a complex pattern of associations between the innate tendency to imitate, the earlier experiences, and recent information about somebody's suicide. The imitative suicide differs from other cases of complex imitation in that it is irreversible (when it results in death). Even when the first suicide attempt is unsuccessful, the person may try to do it again and again. In this case, the repeated attempts may be considered a kind of defensive instrumental response consisting in the use of a specific motor behavior to escape from the aversive sensory state. This often leads to the final act, with no return.

Observations show, however, that only some persons imitate the events that they see or hear about. This suggests that imitation of somebody's behavior requires the presence of a specific sensory state. For instance, suicide imitation occurs mostly in persons whose previous experience and the sensory state are similar to those of the victim (such as the presence of stress or depression).

Inhibition of Imitation

Imitative behavior is observed in many, but not in all subjects. When trying to answer the question as to why a person (actor) does not imitate the behavior of another person (model) we must take into account the actor's previous experience and the resulting sensory state. The results of observations by Bandura et al. (1963) (quoted in chapter 3) suggest that the children who saw the film about the punishment of aggressive behavior restrained their aggressive tendency to avoid unpleasant consequences. Using the terms of conditioning, imitation was inhibited by an avoidance conditioned reflex established by a previous demonstration of punishment for aggressive behavior.

In life, the usual instructions of parents and teachers, the information about the proper behavior required by law, information about negative consequences of crime, and so on, certainly influence the behavior of most members of society. However, there are cases when educational methods seem ineffective. This happens when activation in the sensory system related to the imitative act is higher than the activation related to the inhibitory factors. In such case, the imitative act will be performed anyway.

Conclusions

1. The ability of imitation is already present in the first hours after birth. This ability is thought to be an inborn function, an *unconditioned reflex of imitation*. The arc of this reflex starts with the activation of the corresponding area in the sensory system of the brain and then, through the inborn connections, it reaches the related area in the brain motor system, resulting in the performance of the demonstrated act.

2. In the process of development of the organism, the ability to imitate develops, involving complex eases of behavior such as eating in satiated state (when others eat), following others in manners, fashionable clothes, life style, and so on, and even in commiting suicide.

3. Imitative behavior can be inhibited by factors antagonistic to the particular imitative case. Such factors include knowledge acquired by education, or personal experience about the negative consequences of particular imitative acts (such as sickness or punishment by law). Imitation may also be inhibited by temporary conditions such as oversatiation (resulting in refusal to eat when others are eating) or the presence of an antagonistic activation which is higher than activation related to the imitative act.

4. Injuries in the prefrontal lobes resulted in disinhibition of previously inhibited imitative acts, leading to excessive occurrence of imitative behavior. This fact suggests that prefrontal lobes of the brain may be involved in the process of inhibiting imitative behavior. However, more studies are needed to fully understand the brain mechanisms responsible for both the process of imitation and the process of its inhibition.

References

Baldwin, J.M. (1895, 1906). *Mental Development in the Child and the Race: Methods and Processes.* New York: Macmillan Co.

Bandura, A. Grusec, J.E. and Menlove, F.L. (1966). Observational learning as a function of symbolization and incentive set. *Child Development* 37: 499–506.

Bandura, A., Ross, D. and Ross, S.A. (1963). Vicarious reinforcement and imitative learning. *J. Abnorm. Soc. Psychol.* 67: 601–607.

Bartashunas, C. and Suboski, M.D. (1984). Effects of age of chick on social transmission of pecking preferences from hen to chicks. *Develop. Psychobiol.* 17: 121–127.

Bayroff, A.G. and Lard, E.L. (1944). Experimental social behavior in animals. III. Imitational learning of white rats. *J. Comp. Psychol.* 37: 165–171.

Bond, C.F., Jr. and Titus, L.J. (1983). Social facilitation: LA meta-analysis of 241 studies. *Psychol. Bull.* 94: 265–292.

Bollen, K.A. and Phillips, D.P. (1981). Suicidal motor vehicle fatalities in Detroit: a replication. *Amer. J. Sociol.* 87: 404–412.

Bollen, K.A. and Phillips, D.P. (1982). Imitative suicides: a national study of the effects of television news stories. *Amer. Sociol. Rev.* 47: 802–809.

Brutkowski, S., Konorski, J., Lawicka, W., Stepien, I. and Stepien, L. (1956). The effect of removal of frontal poles of the cerebral cortex on motor conditioned reflexes. *Acta Biol. Exper.* 17: 167–188.

Burd, A.P. and Milewski, A.E. (Cit. by Meltzoff and Moore 1983).

Carroll, W.R. and Bandura, A. (1985). Role of timing of visual monitoring and motor rehearsal in observational learning. *J. Motor Behav.* 17: 269–281.

Carroll, W.R. and Bandura, A. (1987). Translating cognition into action: The role of visual guidance in observational learning. *J. Motor Behav.* 19: 385–398.

Chesler, P. (1969). Maternal influence in learning by observation in kittens. *Science* 166: 901–903.

Church, R.M. (1957). Transmission of learned behavior between rats. *J. Abnorm. Soc. Psychol.* 54: 163–165.

Darby, C.L. and Riopelle, A.J. (1959). Observational learning in the rhesus monkey. *J. Comp. Physiol. Psychol.* 52: 94–98.

Davis, J.M. (1973). Imitation: A review and critique. In: P.P.G. Bateson and P.H. Klopfer, eds., pp.43–72. *Perspectives in Ethology.* New York: Plenum.

Dobrzecka, C. and Wyrwicka, W. (1960). On the direct intercentral connections in the alimentary conditioned reflexes type II. *Bull. Pol. Acad. Sci. CI,* V, 8: 373-375.

Dunkeld, J. (1978). The function of imitation in infancy. (Unpubl. doctoral diss.; quoted by Meltzoff and Moore 1983).

Eisenberg, L. (1986). Does bad news about suicide beget bad news? *N. Engl. J. Med.* 315: 705-707.

Field, T.M., Woodson, R., Greenberg, R. and Cohen, D. (1982). Discrimination and imitation of facial expressions by neonates. *Science* 218: 179-181.

Galef, B.G.,Jr. (1977). Social transmission of food preferences: an adaptation for weaning in rats. *J. Comp. Physiol. Psychol.* 91: 1136-1140.

Galef, B.G.,Jr. and Clark, M.M. (1971). Social factors in the poison avoidance and feeding behavior of wild and domesticated rat pups. *J. Comp. Physiol. Psychol.* 75: 341-357.

Galef, B.G.,Jr. and Clark, M.M. (1972). Mother's milk and adult presence: two factors determining initial dietary selection by weanling rats. *J. Comp. Physiol. Psychol.* 78: 220-225.

Goethe, J.W.,von (1774, 1988). *The Sorrows of the Young Werther.* New York: Hippocrene Books.

Gould, M.S. and Schaffer, D. (1986). The impact of suicide in television movies: Evidence of imitation. *N. England J. Med.* 315: 690-694.

Gould, M.S., Schaffer, D. and Kleinman, M. (1988). The impact of suicide in television movies: replication and commentary. *Suicide and Life-threatening Behavior* 18: 90-99.

Graciano, A.M., DeGiovanni, I.S. and Garcia, K.A. (1979). Behavioral treatment of children's fears: a review. *Psychol. Bull.* 4: 804-830.

Guillaume, P. (1926, 1971). *Imitation in Children* (transl. by E.P.Halperin). Chicago: The Chicago University Press.

Halasz, G. (1987). Self-poisoning and suicide attempts by burning. *Br. J. Psychiat.* 151: 267.

Harlow, H.F. (1932). Social facilitation of feeding. *J. Genet. Psychol.* 41: 211-220.

Hess, E.H. (1973). *Imprinting: Early Experience and the Developmental Psychobiology of Attachment.* New York: Van Nostrand Reinhold.

Heyes, C.M. and Dawson. G.R. (1990). A demonstration of observational learning in rats using a bidirectional control. *Quart. J. Exper. Psychol.* 42 B (1): 59-71.

Hinde, R.A. (1969). *Bird Vocalization in Relation to Current Problems in Biology and Psychology.* Cambridge: Cambridge University Press.

Jacobson, S.W. (1979). Matching behavior in the young infant. *Child Development.* 50: 425-430.

Johanson, I.B. and Hall, W.G. (1981). The ontogeny of feeding in rats. V. Influence of texture, home odor and sibling presence on digestive behavior. *J. Comp. Physiol. Psychol.* 95: 837-847.

John, E.R., Chesler, P., Victor, I. and Bartlett, F. (1968). Observation learning in cats. *Science* 159: 1489-1491.

Kaminer, Y. (1986). Suicidal behavior and contagion among hospitalized adolescents. *N. Engl. J. Med.* 315: 1030.

Katz, D. and Revesz, G. (1921). Experimentelle Studien zur vergleichenden Psychologie. *Z. Angew. Psychol.* 18: 307–320.

Kawamura, Y. Personal observations.

Konorski, J. (1967). *Integrative Activity of the Brain: An Interdisciplinary Approach.* Chicago: University of Chicago Press.

Lhermitte, F., Pillon, B. and Serdaru, M. (1986). Human autonomy and the frontal lobes. Part I: Imitation and utilization behavior. A neuro-psychological study of 75 patients. *Ann.Neurol.* 19: 326–334.

Lorenz, K.A. (1935). Der Kumpan in der Umwelt des Vogels. *J. Ornitologie.* 83: 137–214. Cited by Hess 1973.

McCall, R.B., Parke, R.D., and Kavanough, R.D. (1977). Imitation of live and televised models by children one to three years of age. *Monogr. Soc. Res. Child Develop.* 42 (5, serial No. 173): 1–93.

McDougall, W. (1908). *An Introduction to Social Psychology.* London: Methuen. Reprinted 1963, London: Morrison and Gibb.

Meltzoff, A.N. (1985). Immediate and deferred imitation in fourteeen-and-twenty-four-month-old infants. *Child Development.* 56: 62–72.

Melzoff, A.N. (1988). Infant imitation and memory: Nine-month-olds in immediate and deferred tests. *Child Development.* 59: 217–225.

Meltzoff, A.N. and Moore. M.K. (1977). Imitation of facial and manual gestures by human neonates. *Science* 198: 75–78.

Meltzoff, A.N. and Moore, M.K. (1983). Newborn infants imitate adult facial gestures. *Child Development.* 54: 702–709.

Miller, N.E. and Dollard, J. (1941). *Social Learning and Imitation.* New York: McGraw-Hill.

Miller M.W. and Vogt, B.A. (1984). Direct connections of rat visual cortex with sensory, motor and association cortices. *J. Comp. Neurol.* 226: 184–202.

Miller, S. and Konorski, J. (1928). Sur une forme particuliere des reflexes conditionnels. *C.R. Seanc. Soc. Biol.* (Paris) 99: 1155–58.

Osborn, E.L. (1986). Effects of participant modeling and desensitization on childhood warm water phobia. *Behav. Ther. Exp. Psychiat.* 17: 117–119.

Ost, L.G. (1989). One-session treatment for specific phobias. *Behav. Res. Ther.* 27: 1–7.

Ostroff, R.B., Berends, R.W., Kinson, L. and Oliphant, J. (1985). Adolescent suicide modeled after television movie. *Amer. J. Psychiat.* 142: 989.

Phillips, D.P. (1974). The influence of suggestion on suicide substantive and theoretical implications of the Werther effect. *Amer. Sociol. Rev.* 39: 340–354.

Phillips, D.P. (1979). Suicide, motor vehicle fatalities, and the mass media: Evidence toward a theory of suggestion. *Amer. J. Sociol.* 84: 1150–1174.

Phillips, D.P. and Cartstensen, L.L. (1986). Clustering of teenage suicides after television news stories about suicide. *N. Engl. J. Med.* 315: 685–689.

Piaget, J. (1945, 1962). *Play, Dreams and Imitation in Childhood* (transl. by C. Gattegno and F.M. Hodgson). New York: Norton.

Preyer, W. (1900). Die Seele des Kindes (5te Auflage). Quoted by McDougall (1908), p. 91.

Rabenold, P.P. (1987). Recruitment to food in black vultures: Evidence for following from communal roots. *Animal Behav.* 35: 1775-1785.

Rzoska, J. (1953). Bait shyness, a study in rat behaviour. *Brit. J. Animal Behav.* 1: 128-135.

Schmidtke, A. and Hafner, H. (1988). The Werther effect after television films: New evidence for an old hypothesis. *Psychol. Med.* 18: 665-676.

Skinner, B.F. (1938). *The Behavior of Organisms: An Experimental Analysis.* New York: Appleton-Century.

Solomon, R.L. and Coles. M.R. (1954). A case of failure of generalization of imitation across drives and across situations. *J. Abnorm. Soc. Psychology* 49: 7-13.

Soulairac, A. et Soulairac, M.L. (1954). Effets du groupement sur le comportement alimentaire. *C. R. Soc. Biol.* (Paris) 148: 304-307.

Stepien, I. and Stepien L. (1959). The effect of sensory cortex ablations on instrumental (type II) conditioned reflexes in dogs. *Acta Biol. Exper.* 19: 257-272.

Stimbert, V.E., Schaeffer, R.W. and Grimsley, D.L. (1966). Acquisition of an imitative response in rats. *Psychonom. Sci.* 5: 339-340.

Suboski, M.D. (1984). Stimulus configuration and valence-enhanced pecking by neonatal chicks. *Learning and Motiv.* 15: 118-126.

Suboski, M.D. (1987). Environmental variables and releasing valence transfer in stimulus-directed pecking in chicks. *Behav. Neur. Biol.* 47: 262-274.

Suboski, M.D. (1989). An acquisition of stimulus control over released pecking by hatchling chicks (Gallus gallus). *Canad. J. Psychol.* 43: 431-443.

Suboski, M.D. and Bartashunas, C. (1984). Mechanisms for social transmission of pecking preferences to neonatal chicks. *J. Exp. Psychol: Animal Behav. Proc.* 10: 182-194.

Tarde, G. (1890, 1903). *The Laws of Imitation* (transl. by E.C. Parsons). New York: Henry Holt.

Thorndike, E.L. (1911). *Animal Intelligence.* Darien, CT: Haffner.

Thorpe, W.H. (1961). *Bird Song: The Biology of Vocal Communication and Expression in Birds.* Cambridge: Cambridge University Press.

Thorpe, W.H. (1963). *Learning and Instinct in Animals.* London: Methuen.

Thorpe, W.H. and North, M.E.W. (1965). Origin and significance of the power of vocal imitation: With special reference to the antiphonal singing of birds. *Nature* 208: 219-222.

Tolman, C.W. (1964). Social facilitation of feeding behaviour in the domestic chick. *Animal Behav.* 12: 245-251.

Tolman, C.W. (1968). The varieties of social stimulation in the feeding behaviour of domestic chicks. *Behaviour* 30: 275-286.

Tryon, A.S. and Phillips, K.S. (1986). Promoting imitative play through generalized observational learning in autisticlike children. *J. Abnorm. Child Psychol.* 14: 537-549.

Turner, E.R.A. (1964). Social feeding in birds. *Behaviour* 24: 1–45.

Waite, E.R. (1903). Quoted by Thorpe and North (1965, p. 219).

Wasserman, I.M. (1984). Imitation and suicide: A reexamination of the Werther effect. *Amer. Sociol. Rev.* 49: 427–436.

Williams, J.M.G., Lawton, C., Ellis, S.J., Walsh, S. and Reed, J. (1987). Copycat suicide attempts. *Lancet* 8550 (Vol.II): 102–103.

Wishart, J.G. (1986). Siblings as models in early infant learning. *Child Development* 57: 1232–1240.

Woolsey, C.N. (1958). Organization of somatic sensory and motor areas of the cerebral cortex. In H.F.Harlow and C.N. Woolsey, eds. *Biological and Biochemical Bases of Behavior*, pp.63–81. Madison: The University of Wisconsin Press.

Wyrwicka, W. (1978). Imitation of mother's inappropriate food preference in weanling kittens. *Pavl. J. Biol. Sci.* 13: 55–72.

Wyrwicka, W. (1981). *The Development of Food Preferences*. Springfield, IL: C.C. Thomas.

Wyrwicka, W. (1988). *Brain and Feeding Behavior.* Springfield, IL.: C.C. Thomas.

Wyrwicka, W. (1990). Relationship between imitative behavior and companionship in feeding in cats. *Pavl. J. Biol. Sci.* 23: 38.

Wyrwicka, W. and Chase, M.H. (1972). Eating as an instrumental reaction rewarded by electrical stimulation of the brain. *Physiol. Behav.* 9: 717–720.

Wyrwicka, W. and Clemente, C.D. (1970). Effect of electrical stimulation in VMH on saccharin preference and water intake in cats. *Experientia* 26: 617–619.

Wyrwicka W. and Long, A.M. (1980). Observations on the initiation of eating new food by weanling kittens. *Pavl. J. Biol. Sci.* 15: 115–122.

Wyrwicka, W. and Long, A.M. (1983). The effect of companion on consumption of ethanol solution in cats. *Pavl. J. Biol. Sci.* 18: 49–53.

Name Index

Subject Index